EARTH
UNDER FIRE

OUR PLANET is warming rapidly, and changes are spreading faster than we realize. For ten thousand years the Earth nurtured human life and fostered the development of civilization, but it is now becoming less hospitable. Everyone, in every nation, will feel the effects of climate change.

This profound global change is caused by how we get and use energy. We must overhaul our energy base, and in a very short time, to prevent accelerating climate disruptions. Fortunately, tools already exist to change our sources and curb our waste of energy—a transition that will also benefit many other parts of our lives. What is most lacking, at this point, is the will to change.

Athabasca Glacier, Canada, 1917 and 2005.
AUGUST 2005; 1917 PHOTO BY A. O. WHEELER

Global warming
is causing
glaciers and ice sheets
to melt at an
accelerating rate.

The disruption of ecosystems caused by global warming is affecting all living creatures, such as this polar bear, who has been forced onto dry land in search of food.

NEAR BARROW, ALASKA, AUGUST 2002

Scientists are studying the responses of plants and animals to higher temperatures in all parts of the world.

SCHRANKOGEL MOUNTAIN, AUSTRIA, AUGUST 2004

As the ice melts at
the poles and in high
mountains and the
ocean warms and
expands, low-lying
areas are inundated
by the rising sea.

BHOLA ISLAND, BANGLADESH, JULY 2005

Coastal cities all
over the world will
be affected.

DELRAY BEACH AREA, FLORIDA, MAY 2001

Signs of our changing climate include catastrophic storms, flooding, and drought.

Fossil fuel use for power and transportation is the greatest source of the greenhouse gases that cause global warming.

CARSON, CALIFORNIA, MARCH 2005

Redesigning our cities to include dense public transportation networks and energy-efficient buildings is an important way to conserve energy, and energy conservation is the fastest and cheapest way to cut greenhouse emissions.

We can use energy from
renewable sources such
as the wind and sun to
create a cleaner, safer,
and cooler world.

SOUTH OF ROCKFORD, ILLINOIS, MAY 2004

For my son, Cedar, and his generation, and the generation to follow:
May they see—may they be—the human promise fulfilled.

HOW GLOBAL WARMING
IS CHANGING THE WORLD

EARTH
UNDER FIRE

GARY BRAASCH

WITH AN AFTERWORD BY BILL McKIBBEN

UPDATED EDITION

UNIVERSITY OF CALIFORNIA PRESS
Berkeley Los Angeles London

The publisher gratefully acknowledges the generous contribution to this book provided
by the General Endowment Fund of the University of California Press Foundation.

University of California Press, one of the most distinguished university presses in the United
States, enriches lives around the world by advancing scholarship in the humanities, social sciences,
and natural sciences. Its activities are supported by the UC Press Foundation and by philanthropic
contributions from individuals and institutions. For more information, visit www.ucpress.edu.

University of California Press
Berkeley and Los Angeles, California

University of California Press, Ltd.
London, England

ISBN 978-0-520-26025-2 (pbk. : alk. paper)

The Library of Congress has cataloged an earlier edition of this book as follows:

Library of Congress Cataloging-in-Publication Data

Braasch, Gary.
    Earth under fire : how global warming is changing the world / Gary Braasch with an afterword
by Bill McKibben.
        p.   cm.
    Includes bibliographical references and index.
    ISBN 978-0-520-24438-2 (cloth : alk. paper)
    1. Climatic changes.   2. Global warming.   3. McKibben, Bill.   I. Title.

QC981.8.C5B715   2007
363.738'74—dc22   2007002259

Manufactured in China

18   17   16   15   14   13   12   11   10   09
10   9   8   7   6   5   4   3   2   1

This book was printed using soy inks on acid-free, totally chlorine-free (TCF), recycled Polymax
Matte. Polymax Matte is a product of sustainable forestry practices and contains 20–25% pre-
consumer waste. It meets the minimum requirements of ANSI/NISO Z39.48-1992 (R 1997)
(*Permanence of Paper*).

PHOTO ON PREVIOUS PAGE: Chicago choking under severe heat and air pollution in July 1995,
when 875 people died. This heat wave paled in comparison to the heat wave of August 2003 in
Paris and across Europe, the deadliest since modern civilization began in Europe.

4/10

CONTENTS

PREFACE TO THE 2009 EDITION

Since the first printing of *Earth under Fire* in October 2007, a great deal has occurred to underscore this book's subtitle, *How Global Warming Is Changing the World.*

The United Nations' community of scientists, the Intergovernmental Panel on Climate Change (IPCC), finished its fourth assessment of the climate system, which concluded that warming was "unequivocal," causing increasing changes across the face of the Earth, and almost entirely due to human emission of greenhouse gases. The IPCC investigated actions to limit the pollution and indicated many ways both to adapt to the changes and to reduce their intensity.

Carbon dioxide emissions continued to rise both in quantity and in parts per million in the atmosphere, exceeding the IPCC's highest estimate.

The IPCC received the 2007 Nobel Peace Prize along with former Vice President Al Gore, whose documentary *An Inconvenient Truth* has informed millions of the dangers of rapid global warming. Gore, like environmental leader Lester Brown and climate scientist James Hansen, now advocates very rapid emission cuts and technology change to counteract the steep increase in carbon output.

The Arctic Ocean suffered the most dramatic change as the summer sea ice cover shrank to its smallest extent ever recorded, and the polar bear, whose habitat is the sea ice, was declared "threatened" under the U.S. Endangered Species Act. Several scientists began warning that "dangerous change," which the UN Climate Convention seeks to prevent, is already upon us.

The nations of the world did move closer to an agreement that would include all of them in emission reductions by 2012. Australia joined the Kyoto Protocol after global warming became an issue that turned its national election. China surpassed the United States in carbon dioxide output, but said it would discuss emission reductions under an international plan.

In the United States, the Bush Administration still resisted mandatory limits internationally and nationally, but most states and hundreds of cities were taking action. Schools installed solar roofs and instituted curricula about climate, and "green" products were on shelves from WalMart to the corner grocery.

The price of petroleum nearly doubled, creating a revolution in American transportation habits, forcing automakers to disavow their gas guzzler models, and raising the price of everything made from oil. Biofuels, highly touted early on, were scrutinized for their environmental and food-supply downsides.

The meltdown of banks and investment houses in the mortgage crisis reverberated worldwide and threatened recession. This refocused leaders on rescuing financial institutions and could possibly steal capital and impetus from investments that would help solve the climate dilemma.

It was in the context of these events that Americans elected a new president, Barack Obama. This profound reweaving of our political and social fabric means that the United States now has a leader who, along with most members of Congress, is committed to much stronger action to limit greenhouse gases, encourage change to renewable energy, and cooperate internationally.

This change follows the direction I outline in the epilogue, which bears repeating here: *"Let me state the goal clearly: No policy should be promulgated, no program initiated, no alliance sealed, no purchase made, no machine designed or built, no land use permitted, no product introduced, no law passed, no politician elected unless the action is a step forward to reduction and reversal of the effect of greenhouse gases."*

The challenges are enormous and complex, given the competing—yet interlocking— crises in finance, health care, and Iraq/Afghanistan. The United States and the world are on the brink of understanding how much is at risk as the climate warms. We are learning that climate effects are interconnected with the way we supply our food and goods and health care, the way we govern ourselves, and the way we relate to one another as nations and cultures. We are just beginning to understand that we cannot drill or dig (or shop) our way to a new energy policy, and how much we could gain from changing our energy sources, buildings, and transportation technology.

Most important in determining our response, we have not fully realized how many of our values are affected by the way we get and use energy—ideals and rights like family, nature and creation, security, purity of food and water, freedom, equality, and independence. Expressing and defending our values is key to pulling through the coming years, reducing climate change humanely, and preventing the immense disruptions that unrestrained global warming will unleash.

This updated edition is a guide for this moment, a journalistic narrative and witness of the science, the implications, the technologies, and the human aspects of climate change.

Portland, Oregon
November 2008

INTRODUCTION

This book is a message from many of the places where the effects of rapid climate change are being seen and where scientists are studying what is happening. It is also a report on what these changes mean and what we can do about them.

As a witness to climate change, I have stood in the empty rookeries of displaced Adélie penguins and felt the chill as huge icebergs separated from an ice shelf in Antarctica. I have seen the jagged fronts of receding Greenland glaciers and observed subtle changes on the tundra. I have tracked down Alpine glaciers depicted in 150-year-old images and rephotographed them to show them wasting away. In the woods of eastern North America I have walked among spring wildflowers and watched for migrant songbirds, which are arriving earlier each season than in decades past. Along the coasts I have seen rising tides and heavy storms erode beaches. I have heard the anguish in the voices of native Alaskans as they describe their village being washed away, of Chinese farmers facing famine caused by drought, and of Pacific Islanders driven from their homes by increasingly high tides. Global warming is affecting the whole world, from the tiniest ocean plankton to humans in their cities and the flora and fauna of entire river basins and mountain ranges.

I began making the observations described in this book in 1999, as part of a photographic project I called "World View of Global Warming." Because many popular articles and books on climate change have been based on predictions, which are easily dismissed, I wanted to look at the Earth itself and report on the changes I saw already under way. As a journalist, I wanted to move beyond the raw statistics, the secondhand and political arguments, and talk directly to the scientists who are documenting

OPPOSITE A cavern at the disintegrating edge of the Marr Glacier, Anvers Island, Antarctica, framing an inflatable boat from Palmer Station research base. (JANUARY 2000)

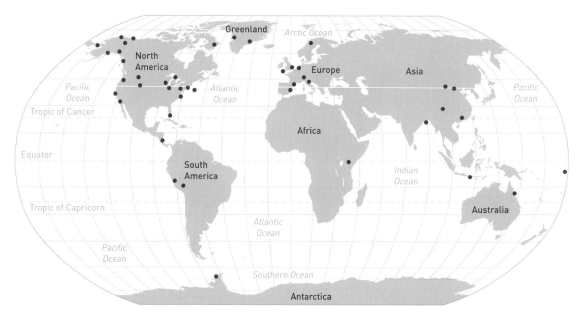

Locations visited in 1999–2006 for the project "World View of Global Warming," which became *Earth under Fire*.

the change. These dedicated and underpaid people, working in remote locations as well as in familiar parks, are devoting their careers to measuring what is happening and interpreting the results. I have crossed both the Antarctic and Arctic Circles, trekked to above 15,000 feet in the Andes, dived on damaged coral reefs, and accompanied scientists into the field in Europe, Asia, Australia, and North America. I raised the money for this personally, year by year, through small grants and assignments, from my photography business, and from the advance royalties on this book. By far the greatest "grant" was one I was given over and over again by individuals—scientists, local people, friends, environmentalists, other journalists—who responded to my requests for information, access, and help.

By now, of course, much more about global warming is known, not only by scientists but by the public and the world's leaders as well. The direct connection between human-made heat-trapping gases and a warming atmosphere is not in question. Hurricane Katrina, the European heat wave of 2003, and increasingly rapid warming in the Arctic, especially, have brought into sharper focus what that warming is doing to our planet. My photographs have been widely used to illustrate these stories in the press and on the web. Because the story is only half told by describing the problem, the project has evolved to show the effect of climate change on people and what they can do about it. In addition to covering some of the ways we get our energy, I look at what we can do to combat near-total dependence on fossil fuel. We already have the tools to change much of this, and a reason to employ them. Many of them are already in use, but we need to do much more. This will require a great deal of will—personal, communal, and national. But taking swift action will produce benefits even beyond lessening the effects of global warming. Although we are rapidly heading toward a different world in any case, we can protect against catastrophe and improve living conditions for plants, animals, and people around the world.

That said, global climate change is advancing far more rapidly than most people realize, affecting the functions of the planet and all our human endeavors. Children born today may see in their lifetimes physical changes to the world that in previous geologic eras occurred over thousands of years. Yet most people, and their leaders, are ill-informed about the multiplying factors that increase the changes and deepen the crisis.

The scientific community has measured, analyzed, and remeasured global warming since early in the twentieth century; thousands of investigations from sky to sea, and from the largest to the smallest creatures on the planet, show its effects. Study by study, report by report, an ensemble of overlapping results strongly and mathematically shows correlations or direct evidence that rising Earth temperatures are having an effect, with feedback cycles magnifying the individual changes. We are altering the physical and biological systems of our world in fundamental ways, ways we would not have thought possible only a few decades ago.

Photographing this subject presents a great challenge. Changes have been unfolding for fifty years or more, with most effects being incremental, or invisible. Pictures are not science; they can, however, provide direct evidence that global warming is happening now, all over the world. They provide contact with eyewitnesses—lifelong observers, Native peoples, and teams of scientists who are seeing rapid change across the expanse of Earth's living systems. Pictures also show that the effects of global warming are taking place in Earth's most beautiful and sensitive landscapes, especially at the extremities of our planet—at the poles, high in the mountains, and in the ocean's rich nearshore environments. Animals and plants on the edge of their ranges, as well as people who live on shorelines or who still subsist from nature, are the first to feel the effects. The photographs in this book are witness to the fact that entire cultures, ecosystems, and species are being forced into transition, their continued existence threatened by our activities.

It is clear to me that I am documenting a decisive, overarching event of the twenty-first century—one with no equal in the previous centuries of human civilization. This is truly a global change. Its consequences might well affect more people than did war in the past century. And no matter how we divide the blame, we are all—all six and a half billion of us—fully at risk. Whether we feel global warming directly or not, we all live with it everyday. We are going to have to adapt to it, as well as to change the things that are causing it. Fortunately, there is a great deal we can do, and many of us have already begun. This is a powerful and urgent story of our civilization that is just beginning to be told.

1

FIRE ON THE ICE

felt like I was exploring a different planet. Four searchlight beams arrowed down from above me to a vanishing point over dark water. Sea fog swept in along the beams and occasionally an iceberg was illuminated. To the left, a small, pale full moon barely showed over the clouds. My exploring vessel was the National Science Foundation research icebreaker *Nathaniel B. Palmer,* a 300-foot floating laboratory whose crew can sample the sea and its creatures to depths of many thousands of feet. I stood on the bridge with the captain and ship scientists, approaching the Antarctic Circle along the west side of the Antarctic Peninsula.

This was my first trip to the southern continent, which I had initially read about more than forty years before as a boy. It was now 1999, and I had just embarked on an odyssey to witness climate change around the globe. In my own lifetime, average temperatures along the Antarctic Peninsula had risen by almost 5°F (2.8°C), according to measurements made at British science stations. This is a huge warming, the greatest known increase anywhere on Earth and, as Antarctic scientists put it in a recent paper, "unprecedented over the last two millennia." The cycles of ice formation and the ocean food chain are being disrupted. Far to the north, the arctic regions and northern, or boreal, forests are estimated to be the warmest in four hundred years, with cascading effects on land and on life. Oceans are warmer, storms are more intense, landscapes are transforming. Nearly everywhere scientists look worldwide, whether they are physically exploring a glacier, analyzing satellite-captured remote-sensing data, or observing changes in animal migration patterns, they see shifts that correlate with rapidly rising temperatures. Scientists studying these effects believe overwhelmingly that almost all recent climate change is being caused by greater amounts of human-made greenhouse gases in the atmosphere. Many of these changes are accelerating: the Earth is becoming a different planet.[1]

TAKING THE TEMPERATURE OF THE POLES

It is hard to accept that the extremities of the Earth, the frozen polar areas, are experiencing the most pronounced effects of global warming. Yet climate models, computer programs simulating the atmosphere's response to changes, foresaw that the Arctic would warm faster than temperate and tropical regions. Antarctica, so large and cold and surrounded by a deep ocean, has been less easy to predict. A wide variety of measurements are being taken, including core samples, land and tropospheric temperatures, radar altimetry, and gravity readings from satellites. Most of Antarctica—namely, the great center of the main, or East Antarctic, ice sheet—is not affected by severe warming at this point. Some researchers believe that Antarctica is somewhat buffered from atmospheric heating by regional winds, the surrounding ocean, and perhaps even by effects of the ozone hole. It is possible that snowfall has increased, which would signify slight warming. However, NASA scientist Robert Bindschadler, speaking at a scientific conference, said the fringes of Antarctic ice are losing elevation all around the conti-

PREVIOUS PAGE The disintegrating face of the Müller Ice Shelf, Lallemand Fjord, Antarctica, 67°S, 2 April 1999. About 5,000 square miles (13,500 sq km) of ice shelves on the Antarctic Peninsula have disintegrated since 1974; because they are already floating, sea level remains constant. However, their absence releases land-based glaciers to debouch directly into the sea, raising sea levels.

GLOBAL WARMING AND
CLIMATE CHANGE EXPLAINED

Global warming refers to an increase in the Earth's average surface air temperature. Global warming and cooling in themselves are not necessarily bad, since the Earth has gone through cycles of temperature change many times in its 4.5 billion years. However, as used today, *global warming* usually means a fast, unnatural increase that is enough to cause the expected climate conditions to change rapidly and often cataclysmically.

Our planet is warmed by radiant energy from the sun that reaches the surface through the atmosphere. As the surface warms, heat energy reflects back toward space; meanwhile, gases in the atmosphere absorb some of this energy and reradiate it near the surface. This is often called the *greenhouse effect,* named for the way heat increases inside a glass enclosure. In the greenhouse effect, the atmosphere can be visualized as a blanket that is made thicker by the action of a small amount of water vapor, carbon dioxide, methane, ozone, nitrous oxide, other gases, and soot; it thus holds in more heat, forcing air temperature higher. The scientific term for this action is, in fact, "forcing."

On an average day, this effect is caused by water vapor and clouds (75 percent) and carbon dioxide (20 percent), with the rest of the heating caused by other gases. Relatively small additions of carbon dioxide and methane force more heat, and that heat allows the air to hold more water vapor, creating a feedback loop that magnifies the effect. Although water vapor is naturally prevalent in the atmosphere, it does not trap as much heat per molecule as carbon dioxide and methane. Also, water vapor molecules cycle through the atmosphere in only a few days, a brief period compared to the residence time of CO_2, which persists for decades and creates some warming even after as long as three hundred years. Dust and aerosol chemicals in the air cause some cooling (negative forcing); they are also very short lived.

Even though the gases are measured only in parts per million (ppm) or billion (ppb), they have been powerfully, and naturally, influencing the Earth's temperature for millions of years. Without them, instead of an average air temperature of about 58°F (14.5°C), the Earth would be below the freezing point. Life as we know it now would be impossible.

Earth's temperature is also subject to natural forcing cycles from solar radiation and the movement of the planet around the sun. Scientists think these cycles, which have left a visible signature extending back millions of years, are what led to past ice ages and the warming that ended them. Currently, we are in a period between major ice ages. The last great glaciation, when temperatures were about 10° to 12°F (6° to 7°C) cooler than today, began fading away about 18,000 years ago. The initial transition out of the ice age was unstable, with many rapid temperature shifts. As temperatures warmed, climate was affected. *Climate* is the accumulation of weather effects—wind, rainfall, heat, cold—experienced in a place over many years, an average of thousands of days' worth of weather. One result of global temperature increase or decrease is *climate change,* referring to a shift in not only average local temperature but also rain- and snowfall, cloudiness and storms, the seasons, and river flow, with associated impacts on the biosphere, the portion of the Earth and its atmosphere that supports life. Although in our daily lives we are attuned to day-by-day swings of temperature and weather, the long-term changes of climate and average Earth temperature are more difficult to apprehend.

During most of the more recent past, the concentration of greenhouse gases remained relatively stable, and so did the Earth's temperature and climate. This was the time when humans developed civilizations and learned how to build cities, grow food, and invent machines. It is possible that early farming and forest clearing had a warming effect on the Earth beginning five thousand to eight thousand years ago. There are also a few examples of natural temperature shifts, such as the Medieval Warm Period, which was followed by the Little Ice Age in the fifteenth through eighteenth centuries. These were possibly not global in extent, and there is scientific disagreement over their causes.

During the Industrial Revolution, people began to use coal and, later, petroleum, to heat cities and run machines. Carbon dioxide in the atmosphere, a by-product of burning both coal and oil, began to increase. Since then, levels of carbon dioxide have risen more than 35 percent, methane concentrations (coming from rice fields, cattle, landfills, and leaks of natural gas) have more than doubled, and nitrous oxide concentrations (another by-product of oil) have gone up by about 15 percent. Some chemicals invented by humans, like chlorofluorocarbons, are also greenhouse gases. Increased greenhouse gases mean more heat is kept in the atmosphere, which led in the late 1800s to a rise in both ocean and air temperature. Between then and 1945, world temperature rose but then leveled off and even decreased a little through the 1960s. The best explanation for that dip appears to be the rise in industrial air pollution, including dust and sulfur, which, as aerosols, cool the atmosphere. The sun's luminosity varied a little through these years, but this appears to have had only minor influence.

Since the 1970s, when laws mandated the reduction of aerosol pollution, atmospheric heat has been rapidly increasing. Whereas the average temperature of the planet rose about 1°F (0.6°C) between the mid-nineteenth century and the end of the twentieth, in the past twenty-five years alone the temperature has risen just over 0.8°F (0.5°C). (The last ice age would have ended in only four hundred years—instead of many thousands—at this rate of heating.) The total heating from the late nineteenth century to 2005 is 1.4°F (0.8°C). The only explanation for these average world readings is the steep rise in greenhouse gases. This rise, moreover, has been shown to be the result not of natural changes but of human activities, primarily the burning of fossil fuels but also farming and forest clearing. Extensive urbanization, air pollution, and increased pumping of water have caused regional change as well. That the added carbon dioxide comes from our actions is certain because this CO_2 has an unmistakable chemical signature.

The greenhouse effect is a matter of natural physics and atmospheric chemistry; it is accurate and uncontroversial to say that higher greenhouse gas concentrations will result in higher temperatures. There is now more carbon dioxide in the atmosphere than at any time in the past 800,000 years, and possibly in the past 20 million years—as we know from the geologic record, in particular ice cores

taken from deep in polar ice sheets. In 2008 the level of CO_2 was above 384 ppm and increasing at a rate of more than 2 ppm per year. Other greenhouse gases are also increasing, bringing the total forcing to about 430 (expressed in ppm of CO_2; about 16.5 billion tons [15 billion mt] of CO_2 put into the atmosphere results in 1 ppm). However, with cooling from aerosols, the actual effect is estimated at about 380. The oceans have absorbed 84 percent of the heating of the Earth system since 1950. That ocean heat, together with the atmospheric heat trapped by added CO_2 and other greenhouse gases, guarantees the Earth will warm even more in coming years. How much more warming will occur is the big question that is discussed in chapter 5.

Scientists are skeptical by nature, and they actively question others' results in addition to subjecting their own work to review by fellow scientists before publication in a journal. In the case of future climate projections, many, many observations and relationships have to be researched and analyzed. Climate is complex, and the result of research is frequently not a single number but a range of probabilities. Naturally, scientists have sometimes disagreed, often strongly, on certain observations or computations. Some of this debate has spilled over into political and public discussion, being taken as "proof" that we don't know very much about climate or that science is evenly divided on the causes of global warming. At this point, no doubt remains as to the big picture, yet questions have persisted in the press and as commentators continue to argue against the scientific evidence.

In fact, science knows a lot about climate change, as this book reveals. The Intergovernmental Panel on Climate Change, an organization of thousands of climate experts that publishes major reports on climate science, impacts, and policy, concludes that the evidence for human-made global warming is strong and consistent, not to mention unusual in the scientific record, and that effects of this change are being seen throughout the natural systems of the planet. World average temperature, according to IPCC projections, will continue to rise anywhere from about 3.2°F to over 7°F (1.8°–4°C) in this century. An unfortunate new finding is that the amount of CO_2 absorbed from the air by the Earth and ocean will decrease, leading to more greenhouse effect. Other factors, like more open ground and water in the Arctic, will also increase the rate of heating. Sea level will rise by 2 feet or more.

On the planetary scale, a shift of even a degree or two is a serious event. The ocean contains about 328 million cubic miles of water, and the atmosphere consists of about 5.5 quadrillion tons of air. The amount of heat or cooling required to shift the inertia of these huge masses is great. But at the same time, the water cycle and biosphere are extraordinarily sensitive to tiny changes. The evidence described and illustrated in this book shows that the delicate balance has started to be upset with little more than 1°F of warming. With temperatures still rising and the greenhouse gas "blanket" increasing in effectiveness, our future is sure to be strongly affected unless we change how we treat the Earth's surface and how we get and use energy. ✦

nent, most rapidly in the west and on the Antarctic Peninsula. This "discharge of glaciers to the sea," he said, "will overcome [the volume added by] increased snowfall."[2]

There is no doubt, though, that West Antarctica and the peninsula are changing. Along the 900-mile (1450 km) peninsula that juts north toward South America, interlocking temperature records estimated from ice cores, ocean sediments, penguin rookery surveys, and thermometer readings show a very steep temperature increase over the last fifty years. The U.S. National Snow and Ice Data Center believes that the breakup since 1974 of seven peninsula ice shelves, comprising 5,200 square miles (13,500 sq km) of ice, is due to this warming. The Larsen Ice Shelf, a sheet of 700-foot-thick (220 m) ice floating out over the Weddell Sea that has probably been stable since the last ice age, has been collapsing into the water in huge chunks since the 1990s. Winters are even more affected than summers. At the British Faraday/Vernadsky Station, for example, the average temperature for the winter months June, July, and August has risen more than 10°F (6°C) since 1951. The extreme warming during winter, when everything normally freezes solid, has reduced not only the extent of Antarctic sea ice but also how long it lasts into the summer. This ice is critical for biological activity, and these changes are affecting penguins and other denizens of the Southern Ocean.[3]

More than 9,000 miles to the north, in the boreal forest and arctic regions, temperatures are also rising, with the consequences that permafrost is warming, precipitation is increasing, and rivers are flowing with more water. The growing season is lengthening on the tundra, and treelines are advancing—yet forests are dying as the incidence of fire and insect attack grows. The Greenland Ice Sheet is thinning at the edges, and the ice cover on the Arctic Ocean is shrinking. All these events correlate with rising temperatures.

The rise in temperature has not been constant. Although Alaska has now warmed to the highest temperature in four hundred years, some places recorded slightly warmer temperatures in the 1930s. Starting in about 1940 much of the area experienced cooling. Since 1966, however, the general picture in the Arctic has been one of warming—at a rate of 0.68°F (0.38°C) per decade, or twice the global rate of temperature rise. Greenland and Scandinavia have shifted from summer cooling to warming, Siberia has experienced more pronounced warming than in the past, and summer warming has continued in the European Russian Arctic. Arctic winter minimums are generally much warmer than the twentieth-century average, as in Antarctica. And nights are warmer—often a more meaningful indicator of change than summer highs. These changes are immense and complicated, affecting the relationships among atmosphere, ocean, and land that propel the Earth's climate. Other layers of complications and interrelationships are seen in the effects on the food chain at both poles (more on this in chapter 2).[4]

ICE SHELVES SHATTER

After crossing the Antarctic Circle, the *Nathaniel B. Palmer* enters Lallemand Fjord and approaches the face of the Müller Ice Shelf, the most northerly ice shelf on the west side of the peninsula. The fjord is packed with ice. Amid the tabular icebergs, which are hundreds of feet long, with sheer cliffs up to 80 feet (25 m) high, are incredible deep blue chunks of glacier ice. The captain carefully steers through a

CERTAINTY AND UNCERTAINTY IN CLIMATE CHANGE

STEPHEN H. SCHNEIDER and
JANICA LANE

Stanford University

Throughout human history, climate has both promoted and constrained human activity. Only very recently have we been able to substantially reduce our vulnerability to natural climate variability, mainly through advances in technology. For example, high-yield agriculture and efficient food distribution and storage systems have virtually eliminated famine in most countries with developed or transitioning economies.

On the other hand, human activity has also affected the climate. For more than sixty years, scientific support has increased for the idea that humans influence nature in complex, and perhaps dangerous, ways, but despite many scientific advances, deep uncertainties about climate change remain—both among scientists and in society. The reasons are many: lack of information, disagreement about what is known or even knowable, linguistic imprecision, statistical variation, scientists' personal biases, and measurement error, among others. The problem is compounded by the global scale of climate change, which has varying impacts at local scales, and by very long-term climate variability that exceeds the length of most instrumental records. It is impossible to have before-the-fact experimental controls or empirical observations; what the climate will be one hundred years from now simply cannot be stated with precision, though qualified estimates can be made.

Climate change is not just a scientific topic, of course; it is also a matter of public and political debate. Stakeholders in that debate may play up or down or otherwise confuse degrees of uncertainty, even as the media pit polarized claims and counterclaims against each other without regard for the relative credibility of each "side."

This does not mean, however, that all aspects of the climate change problem are uncertain. Many of them are well understood. It is scientifically well established, for example, that the Earth's average surface air temperature has warmed significantly—by about 0.7°C, or nearly 1.3°F—since 1860, with a steep upward trend clearly discernible by plotting historical temperatures. Indeed, nineteen of the twenty warmest years on record have occurred since 1980—twelve of them just since 1990. Most of the recent change is attributable to the greenhouse phenomenon, which is solidly grounded in basic science. Greenhouse gases like carbon dioxide and methane have slowly increased naturally since the last ice age waned some 15,000 years ago, changing the balance of solar radiation entering and radiant heat exiting the atmosphere. Water vapor is also a major greenhouse gas, but atmospheric temperature changes appear to be driven primarily by changes in carbon dioxide and methane. In the past few centuries, atmospheric carbon dioxide has increased by more than 30 percent, and virtually all climatologists agree that the cause is human activity, predominantly the burning of fossil fuels and, to a lesser but still considerable extent, land use changes such as deforestation. Methane, mainly a by-product of fossil fuel production and land-use changes such as agriculture, is also increasing rapidly. It is widely accepted that increases in human emissions of carbon dioxide, methane, and other greenhouse gases are largely responsible for recent increases in global average surface temperatures. Even President George W. Bush, an opponent of international mandatory climate policies, has admitted that the warming is real and that humans are to some degree responsible.

Climate scientists believe that the approximately 0.8°C of warming detected since the mid-1800s is only the tip of the iceberg, so to speak, and that warming will likely accelerate in the future. In its Fourth Assessment Report from 2007, the Intergovernmental Panel on Climate Change projected that by 2100 the planet would warm by an additional 3.2°F to over 7°F (1.8°–4°C), and some believe it could warm even more. While warming at the low end of this range would likely be less stressful, it would still be significant for some systems, as many studies are already showing. Warming at the higher end of the IPCC's projected range could have widespread catastrophic consequences.

Based on these temperature projections, the IPCC produced a list of likely effects of climate change. These include more frequent heat waves; more intense storms and a surge in weather-related damage; increased intensity of floods and droughts; warmer surface temperatures, especially at higher latitudes; and rising sea levels, which could inundate coastal areas and small island nations.

Climate change is already happening around the world. We urgently need policies and active measures to slow down the rate at which we are adding to the greenhouse gas burden on the atmosphere. By acting now, we may be able to buy time— to better understand what might happen and to develop more economical carbon mitigation options. Easing the pressure on the climate system is the only insurance policy we have against a number of potentially dangerous and irreversible climatic events. ✦

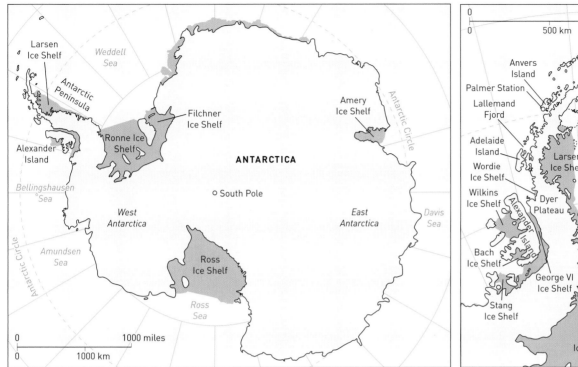

Antarctica and the Antarctic Peninsula. In the map on the right, dotted lines show the boundaries of the Larsen Ice Shelf before large sections broke off in 2002.

slurry of brash ice and gray water, around and between the large bergs. The ship shudders as its engines push ahead to a spot where a year before geologists took measurements in clear blue water right in front of the ice shelf. But now the ice front is still many yards ahead, and the ship cannot get closer, fended off by the frozen debris of disintegration. This small shelf, fed by glaciers from the Loubet Coast, has been receding recently, after steady growth during a four-hundred-year cooling period. Like other receding ice shelves, it may well be a sensitive indicator of rising regional temperatures.

World attention focused on this area in 2002 with the precipitous disintegration of the Larsen Ice Shelf, a 715-foot-thick (220 m) icy webbing between mountainous promontories jutting into the Weddell Sea. Once spanning 5,500 square miles (14,250 sq km) on the eastern side of the peninsula, it began breaking up in great sections in the 1990s. Then, in a thirty-five-day period in early 2002, a section larger than Rhode Island shattered, attracting public notice through a series of time-lapse satellite images released by the National Snow and Ice Data Center (NSIDC). Analyzing the situation, glaciologists determined that the rising temperatures were causing water to pool on the surface of the shelf. As this water percolated down, it weakened the ice along the shelf's vertical faults and fissures, and eventually the 1,250-square-mile (3250 sq km) section collapsed virtually all at once, with hundreds of great vertical slabs of ice falling into the sea and floating away. Since 2002, another 650 square miles (1700 sq km) has broken away.[5]

Geologists believe this portion of the shelf may have been stable for some 12,000 years before the collapse—that is, nearly all the way back to the last ice age. One of the first scientists to figure out the age of the Larsen was Eugene Domack, a geoscientist at Hamilton College in Clinton, New York. His research has involved the analysis of seafloor sediments near the disintegrating front of the Larsen, Müller, and other ice shelves, through which he has gained an understanding of the history of the ice over the past 10,000 years. On the basis of this research, he and his team consider the breakup of the Larsen Ice Shelf an "unprecedented" and "unique" event in recent geologic history, and one that "supports the hypothesis that the current warming trend in the northwestern Weddell Sea has exceeded past warm episodes in both its magnitude and duration."[6]

As falling snow drops a curtain over the blue of Lallemand Fjord, I watch as Domack directs technicians using a large A-frame crane on the stern of the *Nathaniel B. Palmer* to lower 10-foot-long (3 m) square tubes, or "kasten cores," into seafloor mud near the Müller Ice Shelf. The sediment samples thus collected date back some 8,000 years. During an April 1999 cruise, technicians also deployed longer, cylindrical cores in waters nearly 3,300 feet (1000 m) deep, one of which yielded 65 feet (20 m) of sediment—the longest core ever pulled from an Antarctic ice

Satellite view of Larsen Ice Shelf disintegration in 2002. It lost a 1,250-square-mile (3250 sq km) section, called Larsen B, which had been stable for about twelve thousand years, and smaller portions continue to break off. Evidence indicates that temperature increases of over 5°F (2.8°C) over the past fifty years caused water to pool on the ice shelf's surface and erode down into the ice until it was unable to hold together. (NASA)

shelf. When a successful core is winched back on deck, Domack's demeanor changes from quietly serious to animated and energized. The core is immediately trundled onto a lab table on a lower deck, where the crew sets to unscrewing one side of the long steel box. It's like Christmas to them, even though to the rest of us the contents would be more disappointing than a lump of coal. What they uncover is mucky sediment, ranging from raw gray rock and claylike particles to layers of greenish material. The green, a sign of living organisms, indicates that the fjord was once open to sunlight, presumably in a warmer period; the gray muck, in contrast, accumulated beneath a thick ice shelf and has never seen the light of day. While Domack records the physical features and colors of the core, others dissect it, analyzing some portions for magnetic and organic content and preserving others for later study. Based on organic content, cores taken on this cruise confirm a current seventy-year warming trend following four hundred years of cooling. These findings interlock with analyses by glaciologist Ellen Mosley-Thompson of Ohio State University, corroborating her interpretation of long-term records of an ice core drilled from the Dyer Plateau, near the southern end of the Antarctic Peninsula. That site, which provides a 480-year paleoclimatic history of the region, likewise indicates a distinct warming trend beginning in the 1940s, culminating in decades that were the warmest in five hundred years.[7]

As dramatic as the collapse of a great Antarctic ice shelf is, it does not raise the ocean level. Since

Geologist Eugene Domack inspects a kasten core from sediment beneath the Müller Ice Shelf. The muck ranged from raw rock and claylike particles to layers of greenish biologic material, reflecting past times when the fjord was respectively covered by thick ice or open to sunlight in a warmer period. Analysis shows this shelf has been receding recently after growing over a four-hundred-year cooling period. (APRIL 1999)

the ice is already floating, displacing seawater, the level stays constant. Ominously, though, behind the Larsen and other ice shelves, landed glaciers flow down from the mountains of the peninsula and feed ice into the shelves. These streams of ice had actually gotten backed up somewhat, trapped by the great mass of ice in the shelf in front of them. When the Larsen disintegrated in 2002, exposing the peninsula's landed edge, scientists were concerned that the source glaciers might let loose and push ice into the sea—an event that surely would displace seawater and raise ocean levels. Satellite monitoring by the National Snow and Ice Data Center shows that five of these glaciers immediately began flowing faster; within a year they were moving up to eight times more rapidly than before. "If anyone was waiting to find out whether Antarctica would respond quickly to climate warming, I think the answer is yes," NSIDC glaciologist Ted Scambos commented. Glaciers that are still behind existing parts of the Larsen Ice Shelf, however, show no significant changes.[8]

Eric Rignot, a geoscientist with the Jet Propulsion Laboratory who works with satellite data, found that feeder ice streams for other lost ice shelves on both sides of the peninsula have also responded since then. The same trend is evident where glaciers flow directly into the sea rather than into ice shelves. Glaciologists from the British Antarctic Survey and the U.S. Geological Survey (USGS) compared more than two thousand aerial photos and one hundred satellite images of 244 glaciers and found that 87 percent have calved (or separated) off, thinned, or pulled back. "It appears that in recent times this large mixed population of floating and tidewater glaciers has responded synchronously" to a climatic warming, the researchers wrote in *Science*.[9]

Farther to the south, Eric Rignot, American glaciologist Robert Thomas, and colleagues found that large, well-studied ice streams in West Antarctica are flowing 60 percent faster than the rate at which new snow accumulates near their sources. These huge glaciers, such as the Pine Island and the Thwaites, still have ice shelves extending into the Amundsen Sea, but satellite measurements show the shelves are thinning. The scientists wrote in *Science* that "although these glaciers are the fastest in Antarctica, they are likely to flow considerably faster once the ice shelves are removed and glacier retreat proceeds into the deep part of glacier basins." Their statement reflects the fact that, unlike the much larger main dome of ice over eastern sections of the continent, West Antarctic ice is resting in a marine basin up to 3,200 feet (1000 m) below sea level. Ocean water that is now about half a degree above freezing not only erodes the floating ice shelves but also is apparently melting the base of the ice streams where they contact bedrock. This underwater face of the glaciers, called the grounding line, is retreating up to two-thirds of a mile (1 km) a year. The great glacier rivers are responding by speeding up or thinning, an effect that extends many miles into the central part of West Antarctica. Satellite analysis reveals huge channels and reservoirs of meltwater under the glaciers. In papers and public statements, Rignot, Robert Bindschadler, and others have warned that if ocean water reaches the deeper ice within the marine basin, a huge amount of that ice could melt or slide out rapidly. These West Antarctic glaciers of the Amundsen Sea contain enough ice to raise sea level by 4 feet (1.3 m), which, said Rignot, is "comparable to loss of ice from all the world's [mountain] glaciers." This is one area for which published research was in short supply in making forecasts for the United Nations 2007 Intergovernmental Panel on Climate Change (IPCC) reports. One reason, said Bindschadler, is that the computer models do not match what seems to be actually happening. Although the actual IPCC sea level rise numbers are conservative by some scientists' estimations, the panel does warn that continued temperature rise could trigger more rapid ice sheet changes.[10]

IN EARLY 2000 I journeyed to Palmer Station, a U.S. base at 64°S, about halfway down the west side of the peninsula. Here, a handful of blue and tan metal buildings perch on a small point of granite jutting beneath the encircling cliffs of the Marr Ice Piedmont, which covers most of the rest of Anvers Island. This outpost of American science was established by the navy in the 1960s and, like the ship, was named for American sealer Nathaniel B. Palmer, who in 1820 was one of the first humans to see Antarctica. The British Port Lockroy (recently deeded to the Argentines) and Rothera bases are not far away. Scientists from these outposts, looking bulked up in orange survival suits, venture out in Zodiac inflatable boats to observe the weather, ice, abundant seabirds, sea mammals, and tiny plants and animals of the Southern Ocean.

The 180-foot (55 m) ice cliff of the Marr Ice Piedmont has been pulling back from Palmer Station and calving into Arthur Harbor for more than thirty years. Although few precise records of the retreat have been kept, the constant calving of the glacier front and new rock uncovered each year bespeak the change.

On an excursion near the base, station boating manager Ross Hein maneuvered our Zodiac through jagged chunks of glacier ice so crystalline they could have been shards of a gigantic Steuben vase. One large piece rolled over just in front of us, pitching the boat—a reminder of how dangerous even small bergs can be as they melt and shift equilibrium. Soon we motored into a passage between the fractured ice cliff and a tiny glacier-capped island. "Two years ago this wasn't here," said Hein of the new strait. "It was a solid glacier front, and we didn't know this was an island." The coastline of Anvers Island is being redrawn in several places that used to appear as "points" but now are revealed to be islands. Between the 1960s, when the first topographic maps were drawn, and 2000, glacier cliffs behind Palmer Station receded about 1,650 feet (500 m). Since then, the wall of ice across from the base has fallen, opening a direct passage behind Norsel Point and revealing yet another island no one knew had hidden beneath the glacier.

Weather readings are taken year-round, and from these it's clear that temperatures have been increasing more in winter than in summer (an effect predicted by computer climate models). This change not only affects glaciers and ice shelves, which do not refreeze as deeply, but also reduces winter sea ice. Although in summer the sea ice thaws, revealing the rocky edges of the continent and lots of open water, in winter the peninsula becomes locked in thick sea ice and resembles the rest of Antarctica. But whereas fifty years ago heavy winter ice occurred in four out of five years, now it occurs only once in five, according to Palmer Station scientist Bill Fraser. Since 1973, sea ice extent has decreased by more than 20 percent—with momentous results for seabirds and sea mammals, as well as krill and other plankton.[11]

GREENLAND'S ICE FLOWS BACK TO THE PAST

When, over a thousand years ago, Eric the Red sailed west and established settlements along an inviting coast with deep fjords in the North Atlantic, he called the place Greenland. The name may sound like the hype of a real estate developer, but in fact it reflects the relatively warm climate of that era. "For nearly 500 years between A.D. 984 and sometime in the 1400s," notes Jared Diamond in *Collapse*, ". . . two fjord systems supported European civilization's most remote outpost, where Scandinavians 1,500 miles from Norway built a cathedral and churches, wrote in Latin and Old Norse, wielded iron tools, herded farm animals, followed the latest European fashions in clothing—and finally vanished." These settlements were abandoned along with others in Newfoundland. A cooling climate beginning around 1300 had shortened summers, clogged the fjords with ice, and created unbearable hardships. Apparently the five thousand or so Vikings could not adapt, nor were they disposed to learn from the ancient experience of the native Greenlanders, the Inuit. Diamond, a professor of geography at the University of California, Los Angeles, suggests that the dogged societal conservatism and environmental carelessness of the Greenland Norse were as much to blame for their demise as the climate change. They had self-imposed dietary restrictions—raising cattle and sheep but eating little fish—that made no sense in the local setting; they diverted precious income to European-style churches and clothing; they cut down too many trees and ignored erosion. According to Diamond, the collapse of a society is never simple and always messy. And usually, climate change is a major factor, especially when it exacerbates environmental mistakes that are already under way.[12]

The coast of Greenland remained locked in ice until the end of the Little Ice Age. European and American settlements were then reestablished as the Northern Hemisphere warmed again. In the twentieth century, however, rapid atmospheric warming began to cook the Arctic, especially after 1960. Temperatures are warmer now than in the previous four hundred years, at least. Today, the fjords, hills, and cliffs of southern Greenland appear as they might have to Eric the Red. Although this change looks to some like just another upward curve in Earth's roller-coaster weather, according to most Arctic climatologists the current heating is much more rapid than at any time during the previous two thousand years. Mark Serreze of NSIDC says that average Arctic winter temperatures are nearly 3°F (about 1.5°C) warmer than only forty years ago, and some areas have warmed much more. Vast glaciers streaming off the Greenland Ice Sheet are flowing faster than has ever been measured. Their terminuses are either calving off flotillas of icebergs or pulling back up, beyond the ends of fjords. Great lakes of meltwater are appearing on the snow-packed surface of the ice sheet, causing glaciologists to fear that this water may be percolating beneath the ice and lubricating the glaciers to flow even faster. According to the Cooperative Institute for Research in Environmental Sciences, 2005 saw the greatest measured melt of Greenland's ice surface in twenty-seven years of satellite records, with 43 percent of it suffering some thawing. Surveys undertaken since the early 1990s by a low-flying NASA plane showed that many coastal glaciers in Greenland are thinning by up to 130 feet (40 m) a year; they are moving so fast that one can see them flow a hundred feet a day.[13]

To get an idea of what this loss of ice means, I joined the 2001 NASA survey for a firsthand view. Flying at low altitude across southern Greenland's coast and up onto the ice sheet, one is also soaring back in time: right into the last great ice age. From coastal headlands and greening tundra split by steely fjords that become increasingly clogged with ice, the landscape soon changes to exposed rock only recently scoured clean by glaciers. Then the raw leading edge of a glacier comes into view, a dirty gray, crevasse-riddled curve of icy cliffs, looking much like a dam holding back the ice beyond. Under the heat of the air and water that surround it, however, this ice can no longer hold together and is actively calving off. Farther on, the glacier's movement organizes the crevasses into echelons that mark the flow. Along the edges, the high points of previous eras of cold and thicker ice show as the sharp charcoal lines of moraines. And towering over the river of ice are the granite sides of mountains, steepened by eons of glacial scouring, and soaring into magnificent cliffs thousands of feet high. As the plane climbs, other canyons can be seen, hundreds of Yosemites in various stages of melt-out from the ice. The farther up one flies, the deeper the ice becomes, until the only rocks showing are isolated nunataks (peaks surrounded by glacial ice) shouldering above the vast ice sheet. Finally only the glaring surface of the mile-thick ice sheet can be seen in any direction. Viewed like this, Greenland is a magnificent living diorama of glacial history, like something one might see in miniature at a national park visitor center.

When the plane, a turboprop Orion P-3 crammed with radar and laser sensors, reached the smooth top of the ice, it reversed course and flew back down the glacier to the sea. Now my view of Greenland, instead of running time backward, displayed how dramatically, and actively, the ice is shrinking. This is the accurate way to visualize Greenland today—realizing that this crucial remnant of the ice age may be on the verge of a great thaw.

The mission I had joined was a repeat of flights made annually since 1993 by a team headed by Bill Krabill of NASA's Goddard Space Flight Center and glaciologist Robert Thomas. Krabill and his crew are perfecting the art of aerial topographic mapping using LIDAR (laser imaging detection and ranging) technology guided by global positioning systems. Previously, this technology was used to help understand and verify data from instruments sent on probes to Mars. Krabill saw the value of the approach for Earth data, too, and since the early 1990s has led flights over arctic ice sheets, primarily in Green-

Kangerdlugssuaq Gletsjer, on Greenland's remote east coast, shows the signs of rapid flow and thinning: an ice-filled fjord, a sharp calving front, deep crevassing, and a widening trim line along the mountains, as well as steepening tributaries (upper left). In 2005, the main ice stream (right) was flowing twice as fast as when it was first measured in the 1990s, at 8.4 miles (14 km) per year. (MAY 2001)

land, along North American coasts, and over Antarctica, to get baselines for these rapidly changing landscapes.[14]

The P-3 pilot, George Postell, a veteran of navy antisubmarine warfare in the same aircraft, flew at a precise altitude, 1,400 feet above the crevassed ice and jagged mountains, to get repeat measurements from the lasers, radar, and cameras aimed out the bottom of the plane. For safety's sake, he left the automatic altitude warning system on, though he frequently violated its limits trying to follow the curving glaciers and navigate around the nunataks. For a first-time passenger like me, the automated warning voice with its constant commands—"Terrain! Terrain! PULL UP!! PULL UP!!"—was unnerving. Many times I watched as we approached looming cirque walls and found myself desperately thinking the same thing: PULL UP!! PULL UP!! But Postell, on his third mission to Greenland, remained cool, and we successfully completed five 8- to 10-hour flights over selected glaciers along both the east and west coasts.

Glaciologist Thomas has been particularly interested in two very large glaciers that have been measured for many years—at first only by repeat aerial photography or by observation of their terminuses. The Kangerdlugssuaq Gletsjer is on the isolated east coast at about 68°N, feeding down from a magnificent set of mountains that hold the ice sheet back. The upper tributaries sweep down past sharp nunataks at about 9,000 feet (2,750 m) elevation and through narrow passes, eventually combining into a miles-long stream of crevasses and medial moraines more than 4 miles (7 km) wide. We flew up and down this glacier several times, and I could see from the amount of crevassing that this ice is in rapid motion—as glaciers go. In 2001, Thomas confirms, it was losing about 33 feet (10 m) of ice thickness each year and was one of the fastest-moving glaciers in Greenland, flowing about 4.4 miles (7 km) a year. Thomas has remeasured the Kangerdlugssuaq twice since then, and unbelievably it has thinned by as much as 325 feet (100 m) more and is now racing twice as fast, 8.7 miles (14 km) per year. That's more than 5 feet (1.5 m) per hour. This makes it the fastest glacier known on Greenland, bumping the Jakobshavn Isbrae on the western side of the island out of first place. Jakobshavn, which according to recent satellite measurements is moving at a speed of almost 8 miles (12.6 km) per year, has long been known not only for its speed but also for producing the most icebergs of any Arctic glacier. NASA readings show that it is continuing a rapid thinning and retreat up the 4.5-mile-wide (7 km) Ilulissat Fjord and may eventually disconnect from its terminal ice tongue.[15]

Flying between these two great ice streams, we were soon above Greenland's ice sheet. At first it looked featureless and so uniform that I had to strain to focus on it. Then variations appeared on the horizon and flowed under the plane: patches of clear blue ice, fields of crevasses, shadowy depressions, flow patterns, textures like white particleboard and formica. Near the top of the ice I could see the Greenland Ice Project (GRIP) and Greenland Ice Sheet Program (GISP) stations, where in 1992 and 1993 ice cores were extracted from 10,000 feet (more than 3,000 m) of ice.

The bulk of Greenland's ice is a remnant of the great ice age that began to wane 15,000–18,000 years ago. The layers at the bottom, however, flattened by the pressure of two miles of ice, were laid down beginning about 110,000 years ago, as that ice age started. This older ice is about 16°F (−9°C) and frozen to the bedrock. Above these bottom layers in the ice column are records of dozens of rapid changes of temperature and moisture. Called Dansgaard-Oeschger (D-O) events, for the geochemists who first

Jakobshavn Isbrae on the western side of Greenland is moving at more than 7 miles (12 km) a year or 30 meters per day. Jakobshavn has long been known not only for its speed but also for producing the most icebergs of any Arctic glacier, which contributes to both freshening of the Arctic and rising sea level. Greenland's ice sheet surface melted more in 2005 than ever before. (MAY 2001)

identified them, these changes are about 1,500 years apart. One shift at the end of the last D-O event—which geologists call the Younger Dryas period, about 11,500 years ago—appears to have been a temperature rise in Greenland of 15°F (8.3°C) over only a decade (see below). This amazing climate event is visible in other records around the world, from glaciers to deserts, in lesser amounts. From this discovery deep in the ice of Greenland, scientists realized for the first time how rapidly Earth's climate can change, and it is one reason so many of them are concerned about the rate of global warming today.[16]

Greenland as a whole is losing up to 36 cubic miles (150 cu km) of ice per year, according to satellite measurements—more than double the rate of a decade ago. Though snow is still accumulating at the higher elevations, intense surface melting is occurring over more than 40 percent of the ice sheet. Glaciers and their meltwater have been rushing into the sea, primarily in the south and in areas below 6,500 feet (2000 m). Sixty percent of ice loss is from glaciers that terminate on land; the rest arises from tidewater glaciers, some of which are the fastest moving. Yet the snouts of two of the most rapidly surging glaciers on the east coast dramatically slowed in 2006, while thinning and acceleration was detected far up these 25 mile (40 km) ice streams. Thus there remain many questions about the mechanics of melting and glacier flow in Greenland. One idea is that the soles of the glaciers are being lubricated by great amounts of water, now forming into lakes along the edges of the ice sheet and disappearing into the ice down huge holes called moulins. If the water reaches bedrock, glaciologists believe it may reduce friction and speed up the flow of the ice. This feedback loop—a self-increasing cycle— could push flow speeds even faster, according to Thomas. Sooner or later the ice sheet itself could come "uncorked," as is occurring to the glaciers in the Antarctic Peninsula that once fed into the Larsen Ice Shelf. Other researchers have noted a possible analogue: the 1,150 square mile (3000 sq km) ice field in Glacier Bay, Alaska, which was up to a mile (1.5 km) thick during the Little Ice Age but disappeared in only about two hundred years, leaving individual glaciers.

Increased melting in Greenland could result in a sea level rise beyond current predictions. As Thomas puts it, "whatever the mechanism responsible for the observed changes, it is not included in existing models used to forecast ice-sheet responses to climate warming. This means that IPCC forecasts—that ice sheets will have but a small effect on sea level over the next century—must be taken with a very large pinch of salt. If an ice shelf does act rather like a loosely-fitting cork in a tilted bottle of wine and if ice shelves are thinning, then a major research priority is to discover how far the bottle is tilted. In other words, how far inland does a glacier 'feel' the effects of ice-shelf thinning or breakup?" The great flows of the Jakobshavn, Kangerdlugssuaq, and other racing glaciers are adding only a few tenths of a millimeter to sea level each year, and no one can guess yet when or if the outflow will begin an irreversible melt-out. Summing the total of all glaciers around the Arctic that have been measured, European glaciologists Andrey Glazovskiy and Mark Dyurgerov reported in 2006 that Greenland's input of meltwater to the sea has increased dramatically compared to that of other Arctic glaciers. Speaking about both Greenland and West Antarctica, Rignot says simply that "the response of those ice sheets to climate warming will be bigger than predicted."[18]

Another argument that polar ice-melt may proceed faster than predicted comes from comparisons with past interglacial periods—the warm episodes between major planetary glaciations. The

Earth is in an interglacial period now, but it differs considerably from the last one, which occurred about 130,000 years ago. Then there was more open Arctic water, boreal forests extended up to 600 miles (1000 km) farther north, the Greenland Ice Sheet was up to 1,650 feet (500 m) thinner, and sea level was about 13 feet (4 m) higher. Apparently back then the Earth's axis was tilted a bit more, which may have caused the Arctic to warm more, but the carbon dioxide (CO_2) content of the atmosphere was also much lower than it is today. Studies of many other past climate records show that at no time in the past 800,000 years, and perhaps much longer, has the CO_2 concentration been as high as the present 384 parts per million (ppm). Jonathan Overpeck and coworkers, who figured out the temperatures during the last interglacial, calculate that a continued increase in CO_2 levels this century could bring us to a temperature equal to that which existed 130,000 years ago, when sea level rose several meters, fed by Greenland meltwater. Greenland holds enough frozen water to raise sea level by 23 feet (7 m).[19]

THE ARCTIC UNDER SIEGE

Across the entire Arctic, ice is diminishing as air temperatures rise above the long-term normal. The permanent sea ice has thinned to an average of 6 feet (1.8 m), a 40 percent loss since 1958, when initial measurements were taken in secret by the first U.S. nuclear submarine, the *Nautilus*. When the navy declassified early sub data about the Arctic Ocean in the early 1990s, that information was compared with recent soundings. The Arctic sea ice has seldom been without some breaks and leads of open water. But satellite observations have revealed that the summer limit of pack ice is increasingly melting back from the coasts. The extent of sea ice at summer's end has declined by more than 10 percent per decade since 1979, reaching an all-time low of 1.65 million square miles (4.3 million sq km) in 2007, after having reached its second lowest recorded measurement in 2005. NSIDC observations document that the total summer ice lost from 1979 to 2007 is four times the size of Texas.[20]

Satellite view of summer Arctic Ocean sea ice cover, showing a decline of more than 25 percent from 1979 (left) to 2005. In 2007 it shrank another 20 percent. Since 1974 the Arctic Ocean permanent sea ice has also thinned from 9 feet to 6 feet. (NASA)

Mark Serreze and his colleagues believe that heat is causing real damage to this crucial white cover, in both summer and winter. With summer heat bringing about further erosion of multiyear-old core ice, the open surface seawater continues to warm bit by bit; meanwhile, the winter freezing period, shorter by fifteen days each decade, fails to refreeze the seawater as thoroughly as before. According to satellite measurements, the winter extent of freezing was below average in 2004 through 2008. As Serreze put it, "the ice seems to be reaching a tipping point." In 2007 he wrote, "a transition to a seasonally ice-free Arctic Ocean as the system warms seems increasingly certain." Some researchers, of course, do not implicate overall Arctic warming in this ice loss but rather point to a natural cyclical variation in the Arctic climate that has led to stronger westerly winds, which in turn drive the ice out past Greenland, leaving more open water to warm up. Cycles are evident in the year-to-year Arctic temperatures and ice area. Serreze described himself as being a "fence-sitter" on this issue five years ago, "but no more." Great cycles of weather that dominate the north have gone into a less intense stage, he said, "yet the Arctic continues to warm and we continue to lose sea ice."[21]

Even if the winter ice loss measured in 2005 and 2006 turns out to be a cyclical variation, the extent of ice is clearly on a downward trajectory. If the Arctic Ocean becomes mostly ice-free in summers, it will unleash a powerful feedback loop to reinforce the warming. Fresh snow and ice reflect up to 90 percent of incoming solar radiation, while open water *absorbs* that much. With less reflective cover and more dark water, more and more of the sun's energy will be absorbed, heating the water, which will then resist freezing. The ice pack will shrink ever faster.[22]

The physical effects of the warming are visible all over the Arctic. At the northern edge of Ellesmere Island, Canada, one of the largest Arctic ice shelves, the 170-square-mile (440 sq km) Ward Hunt, fractured in August 2002, and it continued to crack into 2005 as the region experienced summer temperatures 5°F (2.8°C) warmer than normal. That August the smaller Ayles Ice Shelf also broke away. The major rivers of Siberia and Eurasia, including the Lena and Ob, are flowing much higher now than seventy years ago, according to geochemist Bruce Peterson of the Woods Hole Research Center and other U.S. and Russian researchers; in 1999 the rivers discharged 7 percent more water than in 1936 into the Arctic, which is the world's smallest ocean.[23]

The mantle of permafrost, the frozen soil that underlies the tundra and boreal forests across a quarter of the Northern Hemisphere, is no longer perpetually frozen. In many places it is thawing. The average permafrost temperature has increased across northern Alaska by up to 7°F (4°C) since 1977, when geophysicist Tom Osterkamp of the University of Alaska began taking readings at thirty sites. The warming is greatest farther north, such as at Deadhorse, in the Prudhoe Bay oil field. In a 2005 paper, Osterkamp writes that although two of his study sites have cooled slightly, readings from the other twenty-eight indicate that "permafrost is warming throughout the region north of the Brooks Range, southward along the transect from the Brooks Range to the Chugach Mountains . . . in Interior Alaska throughout the Tanana River region, and in the region south of the Alaska Range . . . to the Talkeetna Mountains." His colleague Vladimir Romanovsky reports from his Alaskan soil temperature readings that the greatest increases over thirty years, ranging up to 2.7°F (1.5°C), occurred along the route of the Trans-Alaska Pipeline. In northern Russia, permafrost has warmed in some places by 5°F (2.8°C) within two decades.[24]

Geophysicist Tom Osterkamp indicates where ground level was when he installed this temperature probe pipe fifteen years ago near Denali National Park, Alaska. The surface permafrost has been thawing and sinking over this period. A very serious effect of this, especially in Siberia, is burgeoning methane emissions from thawed peat and Arctic lakes. (JULY 1999)

Osterkamp says that the active layer, the surface soil that thaws each summer, appears generally stable in the farthest north. However, in some places in Alaska and Siberia ice wedges within the soil have been thawing, creating lakes. Where lakes already existed, the thaw has been perverse, melting the bottoms of some ponds and allowing the water to drain away and in other cases softening the shorelines so that lakes have enlarged. Farther south in some locations the surface soil does not refreeze year after year and instead slumps into mucky swamps in a process called thermokarsting. Water pools on the surface; riverbanks and hillsides give way; forest trees tilt drunkenly and fall over; human structures and roads cave in; and Native life is disrupted. Many Arctic permafrost soils are rich in plant material, roots, and animal bones, which can decompose rapidly when thawed. Arctic biologists and geologists are concerned that increasing thaw is already releasing much more carbon dioxide and methane into the air than previously suspected. (These effects on biologic and human life are explored in the next chapter.)[25]

"The Arctic is rapidly losing its permanent ice," twenty-one scientists wrote in 2005. "The Arctic system is moving toward a new state that falls outside the envelope . . . that prevailed during recent Earth history. This future Arctic is likely to have dramatically less permanent ice than exists at present. At the present rate of change, a summer ice-free Arctic Ocean within a century is a real possibility, a state not witnessed for at least a million years." The authors, members of the National Science Foundation Arctic System Science Committee, say that the loss of ice appears to be driven largely by "feedback-enhanced global climate warming." They don't think anything can stop it.[26]

THE CURRENT SITUATION IN THE NORTH ATLANTIC

In the Atlantic Ocean, another major change may be looming. The North Atlantic is the source of some of the great ocean currents, crucial to world weather and temperature. The Gulf Stream, for example, brings a huge flow of warm water from the south; it then splits into two currents, one of which continues up past Europe, helping to warm that continent, while the other swings to the west. When that water cools, it becomes denser and sinks, then flows back south in the deep Atlantic. Millions of tons of water per second plunge down, propelling what has been called the "Great Ocean Conveyor," a system of huge currents that transfer heat throughout all the oceans.

The conveyor metaphor was popularized in the late 1980s by geochemist Wallace Broecker of Columbia University's Lamont-Doherty Earth Observatory in a series of articles discussing abrupt climate change. This challenge to the prevailing assumption that Earth's cli-

The Great Ocean Conveyor currents that influence climate worldwide are driven by salinity and temperature differences. When warm Gulf Stream waters reach the sub-Arctic, they give up heat and sink to become a huge cold return flow. An increase of fresh water from melting Arctic ice could slow this process, dramatically altering the climate of the Northern Hemisphere, though so far, only limited changes have been observed. (ILLUSTRATION BY JACK COOK, WHOI)

mate changed only slowly had been brewing among scientists for some time because of evidence found in ocean sediments and new discoveries about ocean circulation. But in the deep ice cores of annual layers of ice from Greenland scientists found abundant evidence of very rapid shifts in temperature, CO_2 concentrations, and atmospheric dust content as the last ice age ended. These are the D-O events. Because the Gulf Stream is so influential in moderating climate, Broecker and others wondered if the rapid changes weren't brought about by this current shutting down, possibly because of pulses of fresh water flowing into the North Atlantic from glaciers to the north and west. Some studies of North Atlantic ocean sediments showed evidence of many icebergs during these times of rapid temperature shifts but nothing that proved the ocean currents actually slowed down.[27]

In 2004, however, scientists at Woods Hole Oceanographic Institution looked farther south, at plankton in core samples taken from the seafloor near Bermuda. These tiny creatures record in their shells subtle changes in ocean chemistry, which in turn are linked to water current speed and temperature. This correlation associated slower Atlantic currents with the coldest intervals as the ice age ended. In this scenario, seawater made much less salty by enormous runoff from the melting continental ice sheet floats on the surface of the North Atlantic, preventing the Gulf Stream water from sinking to complete the cycle of currents. With the cycle interrupted, the warm water flowing from the south slows or stops altogether, and the Northern Hemisphere suddenly gets a lot colder—dropping 8° to 10°F (4.5–5.5°C) in just a few years. The end of such an episode can happen even more quickly, as was seen in the ice core record from the end of the Younger Dryas period.[28]

How this mechanism actually works and how it is triggered remain speculative, but scientists do know that these events happened at the end of periods of deep glaciation when enormous amounts of ice were melting. Wallace Broecker's concern was that now, far removed from any ice age, human-caused warming might be re-creating a similar situation of fresher water blocking the Gulf Stream. To be sure, that current is not the only influence on European temperatures (westerly winds play a large role); nevertheless, climate in Europe and northeastern North America will likely chill—and possibly very quickly—if the Gulf Stream slows dramatically. This conjecture and other evidence of rapid climate change led to a Pentagon study of possible global political and national security disruption following such changes, and, in a grossly speeded up dramatization, to the 2004 summer scare movie *The Day after Tomorrow*.[29]

Could this really happen soon? The answer seems to be no—but it may be getting closer, whether or not humans have any influence over such a huge switch in climate. Ruth Curry at Woods Hole Oceanographic Institution has documented declining salinities in the regions most critical to the Great Ocean Conveyor, whereas more southerly waters near the equator have increased in salinity over the past forty years. During a scientific cruise to study parts of the Gulf Stream in October 2002, Curry explained to me what this means in terms of the Earth's climate. "Something has accelerated the world's water cycle by 5 to 10 percent," she said. The large-scale ocean changes we are seeing suggest a fundamental shift in evaporation and precipitation patterns. As evaporation rates increase in the low latitudes, precipitation in the high northern latitudes rises concomitantly.

To investigate this phenomenon, Curry and Cecilie Mauritzen, from the Norwegian Meteorological Institute, set out to determine the effect freshwater input has had on the North Atlantic since the mid-1960s, using water salinity data collected by hundreds of ships. The ocean is less salty now, due to the huge amount of fresh water estimated to have entered the system—about 4.5 cubic miles (19,000 cu km) between 1965 and 1995 (equivalent to more than three times the annual flow of the Amazon). But the scientists found that "the observed freshening does not yet appear to have substantially altered" currents. "At the observed rate [of input], it would take about a century to accumulate enough fresh water to substantially affect the ocean exchanges . . . ," they wrote, "and nearly two centuries of continuous dilution to stop them."[30]

The ocean system is complex, however, and another study published just a few months later in 2005 showed change in some currents. These direct measurements of the Gulf Stream and some of its return currents by Harry Bryden and colleagues at the National Oceanography Center in Southampton, England, are the first to reveal any slowing of the conveyor. Reporting in *Nature,* they wrote that at latitude 25°N (the line from Florida to just south of the Canary Islands off Africa) less cold water was flowing back southward in the deep return current. Significantly, the source of the Gulf Stream had not slowed at all. Part of the warm northward surface current was peeling off to the east, so that about 30 percent less was reaching the North Atlantic. The specter of less warm water making its way north to keep Europe at a moderate temperature was an obvious concern, but Bryden's subsequent research shows wildly varying measurements that do not lead to any conclusions. The reported change in the currents leaves many questions unanswered, including why Europe has been getting warmer in recent years. Also, Curry pointed out, although "no substantial slowing has been detected in the source currents, the northward flow of warm surface waters, and the southward flows of cold dense waters from the Nordic Seas," measurable freshening has been taking place in the north. What the Bryden study revealed, said Curry, is that the circulation *between* these sources—"analogous to a cog in the ocean conveyor belt"— may be slowing. The new measurements "should be taken as another piece of evidence that global warming is beginning to alter fundamental parts of the climate system—but not as an indication that the ocean circulation is on the verge of a collapse." Although some researchers think the odds of a circulation collapse within two hundred years are "more than two chances in three," the IPCC concluded in 2007 that it will only slow during this century.[31]

New ice core research in Antarctica indicates that more may be involved than North Atlantic water and human greenhouse warming. The temperature fluctuations in the Antarctic ice are the reverse of those in the Greenland cores—not as sharp or as fast but in the same time periods and sequence. It appears that the conveyor somehow links the poles, transferring warmth back and forth over long periods of time, like a seesaw. The last big switch was the Younger Dryas period. The ultimate cause of these oscillations is not known; it could be the conveyor switching on and off. Stefan Rahmstorf, an oceanographer at the Potsdam Institute writing on realclimate.org, thinks they may indicate heat being moved around rather than the whole planet getting warmer and cooler. What humans are doing now is very different, he says, because "a global warming of 3 or 5°C within a century, as we are likely causing in this century unless we change our ways, has so far not been documented in climate history." Because the Arctic and surrounding oceans have a profound influence on the whole planet, the gamble of human influence there seems risky indeed.[32]

IN THE MOUNTAINS, THE END OF AN ICE AGE

Away from the poles, in the mountains, the rest of the world's ice is withering away as well, likewise under assault from rising temperatures. There are more than 160,000 mountain glaciers in forty-one nations, 46,300 of them on the Himalaya-Tibetan plateau alone. Hundreds have been carefully studied and thousands measured by satellite. Except for a very few, they are losing ice mass at an unprecedented and accelerating rate. "The global retreat of mountain glaciers during the twentieth century is striking," write Wilfried Haeberli and Martin Hoelzle of the World Glacier Monitoring Service, University of Zurich; what is more, the trends of shrinking length and volume of ice "represent convincing evidence of fast climatic change at a global scale."[33]

From Indonesia to Alaska and China to Patagonia, most mountain glaciers have been receding since the end of the last cool period on Earth, the Little Ice Age. Toward the end of that cooling in the seventeenth and eighteenth centuries, glaciers advanced far down their valleys, even crushing some villages in the Alps. But since the mid-1800s (with the exception of a minor cooling episode from 1940 to the 1960s that affected some mountain ranges), slowly increasing warmth has overcome this advance. This might be seen merely as a rebound from the cooler period, because many glaciers began their meltback before exceptional amounts of greenhouse gases were being added to the atmosphere. However, glacial recession and thinning are now happening in every mountain range of the world, and the pace has increased—in some areas, the loss of ice is occurring twice as fast as it was thirty years ago.[34]

The wide effects of glacial retreat were well summarized at a 2003 meeting at the National Snow and Ice Data Center in Boulder, Colorado. They include the erosion of recreational options, including tourism and alpinism; opening of new terrain for plant and animal migrations; exposure of new mineral resources; loss of human history and culture, and loss as well of the climate history contained in layers of ice; and ever more catastrophic events, such as the rock-ice avalanche at Kolka Glacier in the Caucasus in September 2002, which killed more than 120 people. The growth of dangerous glacier-dammed lakes is another major threat. A recent survey pointed to forty-four lakes in Nepal and Bhutan, dammed behind moraines, that are rapidly filling with glacial meltwater and threaten to flood valleys and villages below. Recent studies show effects on land uplift, earthquakes, and the Earth's rotation from loss of the weight of glacier ice. Although history is often lost as glaciers shrink, receding glaciers have also brought us face to face with our past—as in Peru, the Alps, and British Columbia, where melting revealed human remains that had been under the ice for hundreds of years or, in the case of Ötzi, a Neolithic man whose body was found near the Austrian-Italian border, five thousand years.[35]

Despite these various concerns, the bigger significance of the loss of this ice held high in the cordilleras of the world can be summed up in two words: fresh water. The Earth's finite supply of H_2O is almost all salty or brackish. Less than 3 percent of water is fresh, and more than two-thirds of that is frozen at the poles. Only a tiny part of Earth's available fresh water is in mountain glaciers, yet, because

Mount Hood, Oregon, 1984. This photo and the next one show Oregon's highest peak, a dormant volcano, draped with its eleven glaciers at summer's end. The ice fields here and on other Cascade peaks are crucial for summer water supply, especially for irrigation in the drier desert plateaus in the rain shadow of the range.

Mount Hood, Oregon, 2002. According to Keith Jackson, Portland State University glaciologist, the ice has shrunk by 34 percent since the beginning of the twentieth century. The Sandy Glacier (center of the photo) is 40 percent smaller.

they are continuously melting, and in turn feeding rivers, especially in summer, these are life-supporting frozen reservoirs. They hold a hundred times more water than is in the world's temperate and tropical rivers. According to the World Food and Agriculture Organization, half the human population drinks water that originates in mountains, and more than a billion people depend directly on flow from glaciers and seasonal snow.

But now global warming jeopardizes every part of this hydrologic system. Snowfall elevations are rising, providing less raw material to low-elevation glaciers. Snow melts earlier and from higher elevations, pushing peak river flows earlier in the year. Runoff from storms and glaciers is already causing floods and pouring more water into the oceans, raising sea level. Rivers are not as full in summer and fall when demand for water intensifies. As glacier ice continues to dwindle, river flows, too, will diminish. Mountain peoples will face drought, as will great lowland cities from China to the American West that depend, at least in part, on glacier water. Many great rivers such as the Ganges and the Rio Grande may run very short of water. As their flow into the ocean becomes erratic and diminishes, creatures as diverse as Bengal tigers and corals will be affected.[36]

The importance of glaciers to so much of the world and the drama of their settings called me to trek into some of the great mountains of the world to try to document what has been lost. In particular, I set out to rephotograph glaciers that had been captured on film years or decades ago. To find my locations, I researched old books and bins of vintage postcards. I pored over topographic maps, trying to locate potentially dramatic views of glacier tongues that would now, I guessed, be shriveled and drawn far up their valleys. In some cases I found out about particular glaciers from scientists or local mountaineers, or ran across photos on the internet. Sometimes all I had to do was drive to a national park, where old photos of huge glaciers often hang in visitor centers. Still, it is difficult to find an earlier photographer's viewpoint even in a park, where roads and trails have been rebuilt and former clearings are now filled in with fifty-year-old trees.

Once I was on a quest to view Austria's longest glacier, the Pasterze, in Hohe Tauern National Park. To reach it I drove through the night in a rainstorm, hoping the storm would break by the time I arrived at the top of the Grossglockner High Road at dawn. I was supremely lucky and got a wonderful view of the glacier's shrunken tongue in the mist below. A few hairpins back down the road, I got even luckier when I spotted a display of four time-series photos that showed the glacier slowly disappearing. As Grossglockner Mountain itself ducked in and out of the storm clouds, I shot my own photo of the granite-sided, forested valley—a valley that in 1875 was entirely filled by the Pasterze Glacier. Recent measurements indicate that this glacier has pulled back 980 feet (300 m) just since the 1980s and is still receding at a rate of more than 60 feet (18 m) per year.[37]

Very dramatic glacier withdrawal has occurred throughout the Alps, in full view of residents, tourists, and scientists. The five thousand Alpine glaciers had already lost 25 percent of their volume in the quarter century up to 2003. The incredible heat of summer 2003 accelerated the recession by a factor of five, and was particularly brutal on higher ice fields. Glaciers across the Alps lost about 10 percent of their volume that year alone, thinning at twice the rate of the previous hottest year, 1998. Temperatures remained above freezing for more than ten days at elevations over 14,600 feet (4500 m), dislodging rocks

Pasterze Glacier, Hohe Tauern National Park, August 2004. Alpine glaciers lost 25 percent of their volume between 1975 and 2000, and another 10 percent in the brutal heat of 2003.

that had been frozen in place for years. Hiking and climbing routes on the Matterhorn and nearby peaks were closed by falling ice and rock, and ninety people had to be rescued. "In the 1990s," glaciologist Frank Paul said in a 2004 interview, glaciers "lost about 70 centimeters a year; in 2003, they lost 3 meters." Permafrost in the European mountains is also thawing, affecting soil stability and undermining roads and buildings. Ski resorts in Switzerland are taking the desperate step of covering more than 130,000 square feet (12,500 sq m) of the Gurschen and Tortin glaciers with insulating PVC-foam tarp to retard melting. But that effort may be futile. Research extrapolating the rate of recent ice losses into the near future predicts Alpine glaciers may be gone within decades. Study leader Michael Zemp of the University of Zurich wrote, "One should start immediately to consider the consequences of such extreme glacier wasting on the hydrological cycles, water management, tourism, and natural hazards."[38]

In North America, glaciers are contracting far up into their highest basins all across the Cascades, Sierra Nevada, and Rocky Mountains. Some British Columbia Coast Range glaciers have receded "to positions they likely have not occupied in the last 8000 years," reported geoscientist Johannes Koch and colleagues. At the same time, the mountains are undergoing an "upward migration of treeline . . . to a level most likely not reached in the last 7000 years. If this trend continues into the future, we may pass through a threshold from the present climate regime into a new, different one, with unexpected consequences for all of us." The effect on the eight hundred mostly very small glacial features in the Sierra Nevada can be judged by representative glaciers carefully measured in 2004, which have shrunk from 31 to 78 percent over the past century. The consequences are particularly poignant in Glacier National Park, which may soon be bereft of its namesakes. The thirty-four named glaciers in this northern Rocky Mountains park have diminished in extent by as much as three quarters since the nineteenth century. Grinnell Glacier, one of the most famous in the park, has shrunk 63 percent and receded about half a mile (1 km) since 1850; it is nearly out of sight of the lake that shares its name. According to USGS glaciologist Dan Fagre, all but a few of the glaciers here will likely be gone by midcentury.[39]

There is a strong connection between some western glaciers and changes in a great sixty-year cycle called the Pacific Decadal Oscillation (PDO). Ice can advance during cool phases of the PDO, then retreats rapidly in warm periods. Although precise data are less available before 1960 or so, it appears that many glaciers receded sharply from about 1915 into the 1940s, then retreat slowed (some glaciers even advanced) until the late seventies, when the current retreat began. There is no question glaciers are now receding like never before as winter snowpack shrinks, but the PDO cycle makes any short-term predictions difficult. This pattern can also be seen in the amount of snowfall and runoff across the American West. However, the influence of rising temperatures is very strong, reports Philip Mote of the University of Washington, and exceeds what can be explained by cyclical variations alone. Members of his Climate Impacts Group at the University of Washington reported that since 1916 there were "unambiguous warming trends across the western U.S. in the cool season" and that other seasonal trends match what is reported globally. "By comparison," they said, "PDO index time series is not well correlated with regional temperature variations, particularly for the [current] period of rapid warming."[40]

TROPICAL GLACIERS

In the Andes, which contains the largest concentration of glaciers in the tropics, melting has been accelerating, driven by a sixty-year temperature increase that is almost twice the world average. I met glaciologist and mountaineer Alcides Ames at his home in Huaraz, Peru, where he handed me a copy of his study on shrinking ice along with two pictures of glaciers to rephotograph. No longer able to guide me himself, he then entrusted me to his American assistant, Bryan Mark. Bryan led me up into the Cordillera Blanca, through snow and knifing winds, to Glacier Uruashraju, at about 15,000 feet (4500 m). We held Ames's 1986 photo in gloved hands, straining to find the landmarks he described for repeating the view. Even though only fifteen years had passed, it looked like a totally different place—the glacier terminus in the photo was totally gone. I scrambled around to take pictures from a number of viewpoints, worried that after the grueling trek I wouldn't get the matched photos. Only after inspecting the transparencies back home did I realize that the glacier had retreated more than a quarter of a mile (0.5 km) in those few years and was just a smudge of white in my photos. Bryan and I had been standing in Ames's footprints, but the glacier was nearly gone. A few days after visiting Uruashraju I trekked alone through the Llanganuco Valley past Mount Huascarán and up onto the shoulder of Huandoy. I held up the second old photo against the mountain landscape again and again, endeavoring to find the photo spot on the ridge. This one had been taken by the Austrian Hans Kinzl during a 1933 expedition. Across the valley, where in Kinzl's image the fat lobe of Broggi Glacier filled a small basin, I could see only twin tarns where the ice had been. This glacier had receded about half a mile (1 km) in sixty-seven years and was now almost invisible among patches of seasonal snow.

Returning to the Callejón de Huaylas valley gave me new appreciation for the deep connection these people have with their highlands. The villagers here, descendants of the Inca, have suffered greatly from these mountains, even as they derive all their water and sustenance from them. In 1970 an earthquake shook loose a portion of the Huascarán glacier, which fell into a steep valley and became a monstrous mud flow, a slurry of slush and ice chunks and dirt and stone. It then jumped a ridge without warning and completely buried the village of Yungay right up to the church steeple, killing twenty thousand. Thousands more died in Huaraz from the quake. The towns were rebuilt—Yungay was moved a few miles away—and life has gone on. But I was struck by the irony, knowing that today, a new, slow-motion disaster is bearing down on this valley, one that will inexorably choke off these people's summer supply of water, threatening them again. Forty percent of the water in the Rio Santa, which drains the cordillera, is from melting glaciers. On the Pacific side of Peru, 80 percent of water resources originate from snow and ice. Andean glaciers that had held their own in the mid-twentieth century have been persistently losing mass since about 1976. The shortage of glacier water in the Andes will soon result in only a summer trickle into the many reservoirs and hydroelectric power stations that serve large cities of the Andes, such as Lima, Quito, and La Paz.[41]

The same type of glaciers that crashed down on Yungay also hold, in tightly compressed layers of the ice, the secrets of their formation over thousands of years. This history of tropical precipitation and climate is rapidly disappearing as these glaciers melt. For many years, tropical areas were a blank spot on the map of world climate, a gap between north, as revealed in Greenland ice records, and south, as

DIMENSIONS OF CLIMATE CHANGE
IN THE MOUNTAINS OF PERU

ALTON C. BYERS

The Mountain Institute

Mountains are elevated landforms with much of their surface in steep slopes. They cover one-fifth of the world's surface and are found in three-quarters of its countries. From base to summit, mountain habitats can range from tropical rainforest to temperate forest, to alpine pasture, to snow and ice, all within a few horizontal kilometers. About one-tenth of the world's people (or more than 600 million) live in mountain regions, and another two billion people depend on mountains for fresh water, clean air, biological diversity, and timber and mineral resources.

Because of their slope, aspect, verticality, mass, and altitude, mountains are particularly sensitive to changes in climate and have been called some of the best natural "barometers" or predictors of global climate change. As noted in the press and scientific literature, the world's glaciers have retreated over the past century, most noticeably in the tropics and subtropics. Rapid melting of snow and ice has formed many high-altitude glacial lakes, increasing the potential for catastrophic downvalley floods. Many alpine plants and animals may be at risk because they are not able to migrate upslope fast enough to cooler, more suitable habitats. Other potential impacts of climate change on mountains include the melting of permafrost, increased risk of debris flows and landslides, erosion from glacial runoff, and depletion of glacier-fed rivers that provide power, agriculture, and drinking water.

Erwin Schneider's 1936 photograph of Queropalca, Peru, rephotographed by Alton C. Byers in 2003.

Little is known about the short- and long-term effects of glacier retreat and warming trends on mountain villages, urban regions, and downstream watersheds. Through the use of a unique set of repeat photographs from the Andes, however, we can examine a range of apparent impacts of climate change on one Peruvian village over the past seven decades and ask to what extent these changes may be linked. Hundreds of landscape panoramas photographed by climber and cartographer Erwin Schneider during expeditions in the 1930s provide a "slice of time" record to study contemporary landscape change in the Husacarán and Huayhuash Cordilleras of Peru.

Since 1984, I have retraced Schneider's footsteps and replicated his panoramas in an attempt to assess mountain land use, forest cover, pasture conditions, glacial recession, alpine disturbance, and any catastrophic events that may have occurred since Schneider's day. In August 2003, I visited the remote village of Queropalca to replicate his 1936 photograph taken from a hilltop above and west of the village. Several surprising changes had taken place in the intervening sixty-seven years.

Schneider's photograph depicts a roadless and very traditional Peruvian village, complete with *ichu* thatched roofs, adobe huts, and a few native *Polylepis* shade trees. No potato fields can be seen on the slopes behind the village, presumably because the climate was too cold for this crop.

A conspicuous change visible in my 2003 photo is Point A, which shows the traces of a 1998 glacial lake outburst that destroyed part of the village to the south. The lake had formed upvalley beneath the Siula Glacier sometime after the 1930s and reportedly broke its ice dam in 1998 in the wake of an avalanche.

Other changes in the 2003 photograph may reflect more positive changes for the village and its residents. Now the slopes are covered with rectangular potato fields (Point B). Upslope migration of agricultural zones has been reported for many regions throughout the Andes in recent decades. Tin roofs have replaced traditional thatch, and use of cement for courts and homes is widespread. Queropalca's population has at least doubled, a road to the outside world and its markets was completed in 2002, and there appear to be more trees. Less positive, perhaps, is the fact that the native *Polylepis* trees have been replaced with exotics such as *Eucalyptus*.

The actual linkages between the numerous variables involved in transforming Queropalca's landscape over the past sixty-seven years remain obscure, as does whether the cumulative change is positive or negative. Yet this example demonstrates the value of an integrated, field-based approach to the study of climate change impacts on people and mountain ecosystems from the Andes to the Himalayas. Research to identify ways that local people and governments can anticipate, manage, and reduce resource-related conflicts associated with climate change could help mountain communities everywhere adapt to an increasingly variable and uncertain environment. ✦

told by Antarctic glacier cores. This void might have remained had not a young geologist been intrigued by an aerial photo of a large sheet of ice in Peru. Lonnie Thompson had intended to be a coal engineer, but a college class turned him toward geology. When he came to Ohio State University (OSU) for graduate work in the early 1970s, he says he "had already dismissed glaciers because they occupy only a small part of the earth where people live, so how could they be important?" Nevertheless, he accepted an invitation to work in the lab at OSU that housed some of the earliest ice cores from Antarctica and Greenland. When he saw the aerial image of Quelccaya Ice Cap in the remote mountains between Cuzco and Lake Titicaca, he wondered if this ice could somehow link the polar cores he was working with. With funding from the National Science Foundation's Polar Program, Thompson and colleagues flew to Peru in 1974 and trudged miles from the nearest road up to Quelccaya, with a summit elevation of 18,600 feet (5670 m). There wasn't enough money for a core drill, so they dug a snow pit by hand. Before them, clearly exposed, was a sharp record of the past few years of snowfall. In 1983 the group returned with a drill and took an ice core to bedrock. Since then, Thompson has mounted fifty expeditions to mountain glaciers in fifteen countries. Although in his book-jammed office this professor with thinning brown hair and glasses does not conform to the popular image of a great mountaineer, Thompson is, at nearly sixty years old, the world's most experienced high glacier researcher. A peerless explorer who has spent more time above 18,000 feet than any living human except perhaps certain residents of Tibet, he has camped on the ice for weeks, monitoring his solar-powered drills as they spin down into the frozen past. He is a man on a mission to get to these shrinking glaciers and extract their history of climate. It is, he has said, like rescuing books from a great library on fire.[42]

This icy "library" is important, says Thompson, because the climate of the tropics affects so much of the world. Half the Earth's land surface is between 30°N and 30°S, and 70 percent of the human population lives in this region. Although most people don't live right next to glaciers, they get much of their water from glaciers and experience the weather patterns that also affect glaciers. In all of Thompson's cores, clear records can be seen of the monsoons and storms that rush off the oceans and up into the peaks, bringing rain and life to the land. Thompson can see the marks of monsoon failures, too, including the drought recorded in eighteenth-century India that killed more than 600,000 people, and droughts that wrecked pre-Incan civilizations in the Andes. The cores also reveal that the tropics felt the slight warming of the Medieval Warm Period and the cooling during the Little Ice Age, several hundred years ago.

The relative youth of these glaciers also tells much about long-term weather and climate patterns. The oldest core date Thompson has found in the tropical Andes is about 25,000 years, at Sajama in Bolivia. On Huascarán above Yungay, the ice was 19,000 years old at bedrock; on Kilimanjaro, 11,700 years. Thomson began to rethink tropical and high-mountain glaciation, because of Dasuopu in Tibet. "I went there to find a very long record. But it's only 8,000 years. Other Tibetan cores came out only 6,000 to 7,000 years, and had plant fragments stuck to the bottom. This is Holocene ice," he exclaimed, using the geologic term for the age in which we live now. Dasuopu is Thompson's highest site, a layer of ice laid down at 23,500 feet (7100 m) in the Himalayas. If it's only 8,000 years old, what does that mean? Why is tropical and high-mountain ice so much younger than the ice in the far north or in the great ice sheet of Antarctica? Thompson also noted that "when glaciers advance in the tropics, they don't advance at the same time as at the poles." The reason, he thinks, is that the tropics are on a different cycle from the 100,000-year cycle that drives the polar ice ages. The tropical timetable is driven by the precession of the Earth's axis—a 21,700-year cycle that, significantly for glaciers, changes the location of the Intertropical Convergence Zone, a low-pressure region near the equator where storms are born. "Where that band is, glaciers are going to grow," said Thompson, while on the other side of the equator the climate is drier and glaciers shrink. Precession explains why most mountain glaciers don't get much older than 22,000 years; it has also meant that there would always be glaciers forming while others shrank. And this, says Thompson, is why "the twentieth century is different. In today's world, no matter where you are, the glaciers are melting."

This accelerating worldwide melting is driving Thompson to return to some of his previous study glaciers to precisely measure their diminishment. At Quelccaya in July 2004 he found preserved clumps of *Distichia muscoides* moss melting out from the sole of the glacier; radiocarbon dating placed them at 3000 B.C. He is mindful of the difficulty of proving these dates beyond doubt. "So many things in science you can question. But when you find wetland plants . . . perfectly preserved; they had to be buried and stay under the ice until now. It's a powerful message," he said. "It has not been warmer there for

Qori Kalis, outlet glacier of Quelccaya Ice Cap, Peru, 1978. This ice fall drains part of the largest ice cap in the tropics, which Lonnie Thompson has been studying since the 1970s. (PHOTOGRAPH © LONNIE THOMPSON)

Qori Kalis Glacier in 2004. The glacier has been retreating almost 660 feet (200 m) per year recently, forty times its rate of change when first measured. Thompson has found mosses exposed by the shrinking ice cap that last were growing about five thousand years ago.

5,000 years." The rapid cooling that buried those mosses in the Andes was worldwide, Thompson believes, matching dates when Ötzi was alive and cold weather caused narrow ancient European oak tree rings, and matching many other ice and sediment core records as well.

In 2002 Thompson made news with his conclusion, based on an ice core and aerial measurement, that the glaciers of Kilimanjaro had decreased 80 percent in the twentieth century. In their report in *Science*, he and his colleagues told a compelling story: "The disappearance of Kilimanjaro's ice fields, expected between 2015 and 2020, will be unprecedented for the Holocene. This will be even more remarkable given that [the mountain's northern ice field] persisted through a severe 300-year drought that so disrupted the course of human endeavors that it is detectable from the historical and archaeological record throughout many areas of the world. A comparison of the chemical and physical properties preserved [in different parts of the summit glaciers] confirms that conditions similar to those of today have not existed in the past 11 millennia. The loss of Kilimanjaro's permanent ice fields," the scientists concluded, "will have both climatological and hydrological implications for local populations who depend on the water generated from the ice fields during the dry seasons and monsoon failures." In a 2006 revisit to the African peak, Thompson found that the ice is retreating 3 feet (almost a meter) a year.[43]

The same fate may befall those who live below what some call the Great World Mountain System, which holds the third largest concentration of ice after Antarctica and Greenland. The Tien Shan, Pamir, Altai, and Himalayas ring Tibet and supply water to nations containing more than 50 percent of the world's population. Research presented at the annual meeting of the National Snow and Ice Data Center in 2003 indicated that over the sixty years up to 1995, Tien Shan glaciers lost about 27 percent of their mass, while from 1952 to 1998 Altai glaciers lost about 10 percent and retreated by 6–25 feet (2–8 m) a year. Loss of ice area has doubled since 1977 in the Tien Shan, compared with the previous thirty-four years. The terminus of Khumbu Glacier, Nepal, the famed route toward the summit of Mount Everest, has receded 3 miles (5 km) from its position in 1953, when Edmund Hillary and Tenzing Norgay set out. Overall, on the 46,300 glaciers in China's western provinces and Tibet, ice has waned 7 percent over the past forty years, according to Yao Tandong, the nation's chief glaciologist. These are average figures, and there are some exceptions, says Lonnie Thompson, yet the four ice cores he has drilled from Tibetan glaciers confirm that the current melting is the most rapid it has been in two thousand years. At the headwaters of the Yangtze, China's largest river, glaciers were retreating at a rate of about 1.7 percent since 1969, and one source glacier pulled back 2,500 feet (750 m) in only thirteen years.[44]

So far this melting has added water to the region's rivers, which include most of the major rivers of Asia, from the Yangtze around to the Indus. In a 2005 report on how climate change is likely to affect water supplies, Scripps Institution climate physicist Tim Barnett and colleagues said, "there is little doubt that melting glaciers provide a key source of water for the region in the summer months: as much as 70% of the summer flow in the Ganges and 50–60% of the flow in other major rivers. In China, 23% of the population lives in the western regions, where glacial melt provides the principal dry season water source." Concern is great, therefore, that the ice will before long be so reduced that these rivers, crucial to so many people for physical and spiritual life, will begin to dry up seasonally. As glaciologist Yao

Glaciologist Lonnie Thompson confronts the receding ice cap of Kilimanjaro in 2000, during his pathbreaking study of this African peak. No new ice has accumulated since 2000, and retreat continues rapidly. From direct measurements, ice cores, and aerial surveys that continued in 2006, Thompson foresees disappearance of the "snows of Kilimanjaro" by the early part of this century. (PHOTOGRAPH BY VLADIMIR MIKHALENKO)

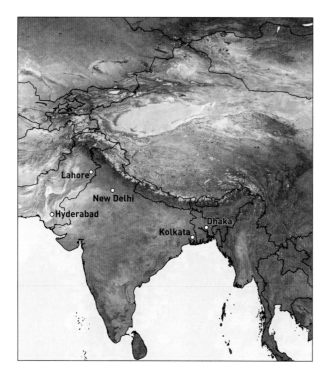

Glaciers in the Himalayas and the Tibetan Plateau. At least 46,000 glaciers, the largest concentration in the world, feed the major rivers of Asia, supporting more than a billion people physically and spiritually in cities such as Lahore, Hyderabad, Delhi, Kolkata, and Dhaka. These glaciers have melted back at least 7 percent over the past forty years. (ADAPTED FROM J. KARGEL AND R. WESSELS/GLIMS/ USGS/NASA)

told the *China Daily,* "The full scale glacier shrinkage in the plateau regions will eventually lead to an ecological catastrophe."[45]

Mountain glaciers, despite their biological and cultural importance, hold only 3 percent of fresh water and cover an area only about the size of France. Yet so much meltwater is cascading down into rivers and seas that it's influencing sea level. Around the globe, the ocean is 6–8 inches (15–20 cm) higher than it was one hundred years ago. Just how much oceans will rise in coming years remains highly speculative, however. In February 2002 glaciologist Mark Meier, reporting on research conducted with Mark Dyurgerov, said that accepted estimates of sea level rise are underestimated because mountain glaciers are retreating more rapidly than anticipated. "The rate of ice loss since 1988 has more than doubled," he said, and by 2100 the contribution to sea level rise from this melting alone will be at least 8 inches (20 cm), and perhaps as much as 18 inches (50 cm). This is in addition to the sea level rise from melting ice sheets in Greenland and Antarctica, plus that caused by expansion of warming seawater. Meier and Dyurgerov, among the world's preeminent glaciologists, say the contribution by land glaciers to the sea has already increased to 1 mm a year, equivalent to 4 inches per century. "Glacier ice melt is accelerating in recent years," they wrote in 2005, "and is likely to continue at a high rate into the future."[46]

Although others, having recalculated the melt rates of mountain glaciers, suggest that total meltwater will add much less than 8 inches, Eric Rignot's measurements at both poles and in Patagonia lead him to agree with Meier and Dyurgerov. "At one point," he wrote me in an email, "a study claimed all was explained by thermal expansion. It is now clearly recognized that this is not correct and that in fact melting ice is the largest contributor." The 2007 IPCC assessment puts the range of sea level rise by the end of the century at 1–2 feet (30–60 cm). But this figure does not account for the meltwater from ice sheets, and the IPCC warns that "larger values cannot be excluded."[47]

Hailuogou Glacier in western
Sichuan, China, the longest glacier
on Mount Gongga in the Hengduan
Mountains. This glacier retreated
about 600 feet (183 m) in the 1990s,
in synchrony with thousands of
other glaciers that flow from the
Tibetan Plateau and Chinese
mountains. (JANUARY 2005)

Portage Glacier, near Anchorage, Alaska, photographed from an adjoining peak during a 1914 geodetic survey. This has been a premier visitor attraction for many years, and in 1986 the U.S. Forest Service built a visitor center on one of the terminal moraines.
(NOAA PHOTO LIBRARY)

Portage Glacier, Alaska, 2004. The glacier has steadily retreated across the lake and by the turn of the twenty-first century had pulled back far into the valley. The ice is no longer visible from the visitor center.

Pasterze Glacier, Austria, 1875, in a photograph from the University of Salzburg glaciology archives. The Pasterze is the largest glacier in Austria and has been carefully measured since just after this photo was made. Above is the highest peak in Austria, the Grossglockner, 12,300 feet (3798 m). (PHOTOGRAPH BY G. JÄGERMAY, COLLECTION OF H. SLUPETZKY)

The same location, rephotographed by Gary Braasch in 2004. The glacier has pulled back out of this deep valley and over the ledge out of sight, a total of more than 3,900 feet (1200 m).

Rhone Glacier, Switzerland, seen from the trail approaching Gletsch, in an 1859 etching by Eugene Ciceri (apparently from a photograph by F. von Martens). This tiny village's location was established—and its name suggested—by the need for travelers heading down the Rhone Valley to cross below the glacier snout. [ETCHING COURTESY STEFAN WAGNER]

Over the years the busy route was improved into a fast highway. The village remained little changed, but the glacier pulled away, revealing the outwash plain and ice fall cliffs. Rhone Glacier is now in its upper basin a mile and a half (2.5 km) distant and 1,400 feet (450 m) higher. [SEPTEMBER 2001]

As we saw earlier in this chapter, there is plenty of evidence to suggest that the seas could rise much higher. "The big gorillas in terms of sea level are Greenland and Antarctica," says Rignot. "The response of those ice sheets to climate warming will be bigger than predicted." One study showed that ice melt could increase sea level up to two meters by 2100. Thermal expansion of the warming oceans will take place at the same time, even if the world temperature rise is halted. Sea level rise is in our future for a long time to come (the ramifications of which are taken up in chapter 4).[48]

Right now polar and mountain landscapes are under fire from warming climate regimes. Retreating ice exposes more land; permafrost thaws; land slides and rises. Entire ecosystems are reacting. North Pole cultures face dislocation and loss of identity while an open Arctic Ocean shifts geopolitics. Inhabitants of alpine regions will have to adapt their traditions, agriculture, infrastructure, and recreation. Glacier-fed rivers will change, as will all who live in these watersheds.

The Earth is becoming a different planet as the ice withdraws.

2

POLAR THAW

It's a blustery day on Torgersen Island, a tiny speck of rock tucked into a bight at the south end of Anvers Island, Antarctica. Even though it is summer here, squalls blow down across the Marr Glacier, pasting snow against the north side of sharp shale promontories. Lost in the wind is the sound of several large Adélie penguin colonies, where tiny chicks are hatching out daily. The weather makes it a rather normal day so far for three humans scrambling up shoreline rocks from the black Zodiac that brought them the half-mile from Palmer Station. Buffeted by gusts and snow, they seek a less windy spot behind a boulder and wriggle out of their thick survival jackets, which they leave behind, weighted down with rocks. They then move up a cobbled plateau past sleeping elephant seals and into a raucous penguin colony. Ecologist Bill Fraser and his Montana State University field assistants are here—as they are every day, weather permitting, from November to April—to monitor the breeding success of penguins caught on the edge of their habitat by a climate shift. Here, as in the Arctic, life, from penguins to polar bears to people, is trying to keep up with the rapid and persistent changes that compromise the ice and tundra.

THE CASE OF THE MISSING PENGUINS

Torgersen Island is one of eleven islets off Anvers Island where thousands of Adélie penguins *(Pygoscelis adeliae)* have nested for some six hundred years. Over the past thirty years this island has lost two-thirds of its eight thousand nesting pairs; other nearby Adélie colonies have declined as well. Bill Fraser, one of the world's senior Antarctic seabird researchers, has documented this change firsthand, having counted nearly every one of these birds each year since 1975. He and the more than sixty graduate assistants and researchers who have participated over the decades have banded more than 19,000 Adélies, and they maintain long-term observations of about 1,000 individual birds. Fraser does not think pathogenic impacts from human contact have caused the drop in population. Rather, he thinks the penguins are decreasing in number because there is both less ice and more snow, a factor he calls "the landscape effect." Fraser believes climate change is to blame.

I accompanied him on a research round in January 2000. Fraser, tall, lanky and weathered, moved among the Adélies, silently counting, noting the extent of each rookery and assessing the success of individual breeding pairs. He directed me down to the leeward, eastern edge of the island, a smooth, pebbly plain that was once the site of a large colony. It's easy to see the former extent of the breeding area, yet only a few pairs of birds were nesting here that year. The breeding success of a pair of Adélies, he explained, seems to correlate quite strictly with where the birds are nesting on the landscape. Rising winter temperatures and diminishing sea ice means there is more water vapor in the air, which causes more snow to fall, accumulating in deep drifts on the lee of these islands. Adélies, deeply loyal to habitual nesting grounds near good feeding areas, lay eggs on the snow anyway, only to have the nests fail because the nest sites are flooded as temperatures increasingly rise above freezing. Eventually, fewer

Ornithologist Bill Fraser stands on the smooth, pebbly nesting ground of what was formerly a large colony of Adélie penguins. He believes penguin nesting here failed because of a decline in sea ice (under which penguins find krill to feed chicks) and more snow from warming temperatures. (JANUARY 2000)

Krill, staple of the Antarctic food chain, may be in shorter supply for predators, including whales and penguins, as a result of warming along the peninsula. A reduction in winter sea ice due to rising temperatures means less summer ice, under which krill feed and develop.
(JANUARY 2000)

chicks hatch and survive to return to their ill-fated birthplace. On the windward side, most colonies are declining too, but not as quickly as on the lee side. More successful colonies take advantage of higher, windward sites where it is safer to nest—although even here the birds are threatened by a recent jump in the population of elephant and fur seals that can steamroll nests and eggs when they move ashore to sleep. Fraser's scientific demonstration that different breeding habitats on a single island promote varied nesting success is helping me to understand the complex effects of rapid warming in the region.[1]

The story does not stop there, however. In the mid-1990s Fraser and other scientists began to study the rapid movement of chinstrap penguins *(Pygoscelis antarctica)* into Adélie territory. Investigations of 600-year-old rookeries using archaeological methods indicate that chinstraps probably did not arrive in this area until about fifty years ago, when average temperatures began to rise. The most common penguins in Antarctica proper, chinstraps are more flexible than Adélies in choosing nesting sites and are able to forage in open water and at night for a wider variety of prey. (Gentoo penguins *[P. papua]*, though less common, have also been moving into Adélie territory on Biscoe Point, Anvers Island, where a receding glacier has opened up much more land.) This population shift is visible even within rookeries, where a distinct line divides Adélies from chinstraps. Fraser described the shift as similar to the successful adaptation of English sparrows, crows, and starlings in American urban areas no longer favorable to native songbirds.

Clues to this phenomenon came from other Adélie rookeries. There are about two and a half million nesting pairs in the whole of Antarctica, and the birds are by no means endangered. Their nesting difficulties seem confined to the far northern sections of the Antarctic Peninsula, where temperature change has been rapid and summer sea ice is diminishing. In the Ross Sea and elsewhere, in contrast, warming has apparently led to an increase in their numbers, as more fractures in the thicker sea ice nearer the South Pole create good access to feeding and breeding locations. The balance seems to hinge not only on suitable nesting sites but also on the penguins' ability to find their prime food, krill, something that turns out also to be strongly affected by climate change. Another Antarctic seabird scientist, Steven Emslie, has likened the situation to a natural experiment in adaptation to rapidly changing conditions.[2]

Krill *(Euphasia superba),* which has perhaps the greatest biomass of any creature on Earth, is the staple food for the entire range of Southern Ocean predators, from birds to whales, including Adélie penguins. Growing to no more than 2.5 inches (6.5 cm) in length, the shrimplike animal has staggering productivity, aggregating in swarms of as many as 30,000 per cubic meter of seawater. Krill prefer to feed on algae and diatom blooms under the sea-ice edge—which explains why less ice makes life hard on Adélies. In years of low sea ice there are fewer krill, causing Adélies, especially juveniles, to forage farther and longer. Fraser, who has fitted some birds with tiny radios, has documented twenty-four-hour feeding journeys, three or four times as long as under optimum conditions. An international team of scientists compared data going back to 1926 and saw that krill numbers have fallen by as much as 75 percent in the Southern Ocean near the Antarctic Peninsula. This 2004 study confirmed that annual krill densities correlate to the amount of sea-ice cover during the previous winter. Other observations implicate periods of reduced sea ice on the eastern Antarctic coast in population crashes and delayed nesting among the well-studied emperor penguins of Terre Adélie (of *March of the Penguins* fame). Antarctic scientists have expressed concern that the link between climate warming, sea ice, and the basic food source, although not conclusively documented, is a powerful sign of how sensitive the continent is to climate change.[3]

GREAT THAW ACROSS THE NORTH

In the north, as with the interrelationships found in Antarctica among birds, ice, and temperature, potentially catastrophic feedback loops are coming into play all around the Arctic Circle. The Arctic Climate Impact Assessment warns that as sea ice shrinks, life zones and species distributions, including those of marine mammals and pelagic birds, are already being affected. Insect attacks on forests and forest fires are on the rise. One of the most pervasive changes is the thawing of permafrost under coasts, tundra, wetlands, and human settlements.

Permafrost covers a quarter of the Northern Hemisphere, and its frozen soils contain about 1,100 billion tons (1,000 billion mt) of dead and decayed organic material. By one estimate, this is more organic carbon than exists in all the living vegetation of the Earth. As the northern climate warms dramatically, more and more of these soils are released from cold storage and decomposition begins, releasing clouds of CO_2 and methane. Methane (CH_4), which has twenty-one times more global warming potential per molecule than carbon dioxide, is generated by decomposition in the absence of oxygen, like in soggy tundra soils. Sergey Zimov, longtime researcher at the Cherskii Science Station on the Kolyma River in northeastern Siberia, working with American colleagues, estimated in 2006 that most decomposition in that area could happen "within a century—a striking contrast to the preservation of carbon for tens of thousand of years when frozen in permafrost." One of the more severe signs in the Kolyma drainage is methane bubbling from thawing muck under expanding lakes, which has increased 58 percent since 1974. To the west, where frozen peat dominates the Siberian landscape, a study revealed that thawed wetlands expel carbon directly into rivers and the Arctic Ocean, where it escapes into the air as CO_2.[4]

A huge swath of tundra and permafrost exists far above the Arctic Circle near Barrow, Alaska, at 71°N. Here in midsummer, the sun never sets, the ground is free of snow, and plants are growing rapidly in the active layer of soil that has melted out above the permafrost. When it is not stormy or foggy, the tundra bakes. Although the air may be cold, it is possible to lie down on the soft lichen and moss in shorts and a T-shirt and be perfectly comfortable. Heat waves ripple up, suggesting the invisible flow of carbon dioxide and methane being exchanged by tundra plants and soils. The average temperature at Barrow is 2.5°F (1.4°C) warmer now than forty years ago, and snow is melting off the tundra on average ten days earlier—though in 2002 through 2004 the process began a whole month earlier. This has made the scientific life a little tougher for ecological physiologist Walt Oechel of San Diego State University. When he started studying the tundra at Barrow in the 1970s, "ten weeks of work would do it" to cover the growing season, from 20 June to the end of August. Now his studies start the fourth week in May and continue into early September. And this trend is accelerating. One recent study reported that the melt season lengthened ten to seventeen days across the Arctic and that recent temperature increases were eight times more rapid than the hundred-year average, near 1°F (0.48°C) per decade.[5]

Oechel works out of the scientific center that has grown up at the east end of Barrow, anchored by the Climate Monitoring Lab of the National Oceanic and Atmospheric Administration (NOAA), where carbon dioxide and ozone are measured continuously. The readings are compared with those from Hawaii, Antarctica, the coast of California, and American Samoa to track the yearly increase of CO_2 and other greenhouse gases in the atmosphere. Nearby, Oechel's Global Change Research Group measures not just the amount but the flux of carbon dioxide—the actual flow of the gas toward the ground as plants take it up, and the flow out of the soil as decomposition occurs. This team was the first to measure these effects in real time on the tundra, using methods developed at NOAA's Atmospheric Turbulence and Diffusion Lab in Oak Ridge, Tennessee. Similar measurements are now being made at more than 270 locations globally. The researchers want to discover how biological and physical processes interact to control carbon uptake, storage, and release in Arctic tundra ecosystems that are undergoing intense heating.

Apart from what is happening in Arctic wetlands, the basic question is this: Is the open tundra, which holds so much of the world's soil carbon, soaking up CO_2, or is it giving it off, thereby adding to all the emissions from other sources? The natural cycle involves high activity during the sunlit summer nights, when tundra plants absorb a lot of carbon, with a near balance of intake and emissions the rest of the year. During the past thirty-five years or so, however, Oechel has seen the tundra shift from being a sink for greenhouse gases—where plant material takes up and stores carbon dioxide—to, sometime in the early 1980s, becoming a producer of CO_2. According to current readings, though, areas around Barrow seem to be adapting in some way during the summer, absorbing more CO_2 again. But the tundra at Barrow remains a net emitter of greenhouse gases. One of the reasons for this great shift and adaptation appears to be a change to drier conditions and a lower water table. This has changed the plant composition. In 1973 the area was primarily wet tundra, but it has been taken over by more dry-site vegetation—coarser sedges and arctic shrubs. Surrounding the North Pole, more areas are now sources of greenhouse gases than are storehouses for them. "The Arctic was the first ecosystem where we saw a complete ecosystem response to climate change," said Oechel. "It's not just local. It is huge."[6]

Southeast of Barrow, near where the Haul Road to Prudhoe Bay crosses the Brooks Range, biologists Gus Shaver and Terry Chapin have seen this transformation in tundra plants take place right before their eyes. Twenty-five years ago, the two scientists and a team of researchers from the Marine Biological Laboratory at Woods Hole, Massachusetts, and the University of Alaska began to monitor vegetation growth near Toolik Lake. To simulate an increase in the Earth's greenhouse effect, they added experimental fertilization and small greenhouses to some study sites. Inside these test plots, dwarf birch trees grew at the expense of sedges, forbs, and other plants that caribou and other wildlife favor as food sources. Even though yearly plant growth doubled, decomposition in the soil speeded up even more—leaving the study areas with less stored carbon than before. Just as in Barrow, these experiments showed tundra was apparently releasing carbon as an effect of global warming. During the research project, which ran from 1981 to 2001 and coincided with the warmest decade in Alaska to that point, Shaver said he even noticed growth changes in his control plots. Abundant leaf litter from the birches was chang-

Biologists inventory tundra plants in experimental plots at Toolik Lake, Alaska. In some long-monitored control plots the sedges preferred by caribou decreased by 30 percent while birch biomass increased, changes correlated to sharply rising temperatures. [JULY 1999]

ing the ecology of the tundra in some plots by killing off mosses. Similar experiments at eleven tundra locations from China to Norway confirm a shift toward shrubs and grasses, diminished cover of mosses and lichens, and a decrease in biodiversity. Comparison of mid-twentieth-century photos with present-day conditions tells a comparable story.[7]

ARCTIC WILDLIFE THREATENED BY A SHIFTING LANDSCAPE

Along the Arctic Ocean shore, the warming is beginning to wreak havoc with the evolved interplay among ocean animals and the birds and mammals that feed on them. Ornithologist George Divoky, for example, has documented the climate-mediated rise and decline of a colony of black guillemots on a barrier island near Barrow. The birds normally nest in cliff cavities, but in the 1970s a few were attracted to crates and barrels left behind at a former navy station on Cooper Island. Their odd behavior succeeded because of the increasing length of summer weather and the slow retreat of Arctic sea ice from around the island. According to Divoky's observations, more than 225 pairs were nesting on Cooper by the late 1990s. But the ice has continued to move away from the island, and with it the prime prey of guillemots: Arctic cod. The birds had to turn to alternate prey. Competition from horned puffins increased. As a result, many guillemot chicks starved or were killed in nest disputes, and the seabird colony declined to fewer than 150 nests at one point, though chick success rebounded in 2005 with slightly better ice conditions. Working twenty-hour days in the midsummer light to check the nests, Divoky and his two assistants have a clear view across the flat, sandy island. In August 2002, he was not surprised to see a polar bear and two cubs wander down from the east end, followed by another that strolled right through the camp. As the summer continued, Divoky had to stop his work on Cooper Island a little early to avoid the foraging bears.[8]

There are about 22,000 polar bears throughout the Arctic. They are the largest of land predators, but they don't belong on solid ground. They really are non-aquatic marine mammals who hunt through the Arctic pack ice offshore and den on and at the edge of the sea ice; hence their scientific name, *Ursus maritimus*. Faced with persistent broken ice or open water that pre-

George Divoky has documented the climate-mediated rise and decline of a colony of black guillemots on a barrier island in the Arctic Ocean. More than 225 pairs nested on Cooper Island in the late 1990s. But the Arctic sea ice has retreated from the island and with it the prime prey of the guillemots, Arctic cod. The seabird colony has declined to fewer than 150 nests. (AUGUST 2002)

vents them from hunting seals and other prey on the pack ice, they can be forced onto land—and into unfortunate contact with humans. The large number of polar bears on land in the Barrow area in 2002 reflects the animal's status as a key indicator of climate change in the north. For many years, adverse effects were reported only at the southern extremes of their range, such as near Hudson's Bay, Canada. That now seems to be changing.

Steven Amstrup of the U.S. Geological Survey's (USGS) Alaska Science Center may know more about Alaskan polar bears than anyone. In more than twenty years of fieldwork, Amstrup has radio-collared some two hundred female bears and then gone on to locate them thousands of times, including in dens along the entire length of the Arctic coast. About half, he says, den up on land or land-fast ice, while the rest spend the winter gestation on sea ice up to a hundred miles offshore. Male bears may den up too, but they also hunt through the winter. Normally, they would be after seals. But early in 2004, surveying in the Colville River delta, Amstrup, fellow polar bear specialist Ian Stirling of the Canadian Wildlife Service, and other scientists came on a scene of cannibalism. A male had apparently smelled a female in her den, crashed in through the roof, pinned her down, and killed her. He then dragged the carcass out onto the ice and partially devoured her. Two cubs in the den had been buried by the caved-in roof and suffocated. Later that spring, Amstrup and Stirling found evidence of two other cannibal-istic attacks, on Canada's Herschel Island. In Amstrup's twenty-four years in Alaska and Stirling's thirty-four in Canada, neither had ever seen evidence of polar bears stalking, killing, and eating an-other bear. These incidents, they wrote in *Polar Biology,* may have resulted from "nutritional stresses related to the longer ice-free seasons that have occurred in the Beaufort Sea in recent years." Amstrup also reported finding three polar bears in 2006 that had "apparently starved to death," a first in his ex-perience in Alaska.[9]

In September 2004, an aerial whale-survey team north of Alaska spotted four drowned bears more than a mile from shore. The team later reported at a scientific meeting in December 2005 "that on the order of 40 polar bears may have been swimming and that many of those probably drowned as a re-sult of rough seas caused by high winds." The observers said that never before in nineteen years of sur-veys had so many bears been seen swimming far from ice, nor had any drownings been reported. "We suggest that drowning-related deaths of polar bears may increase in the future if the observed trend of regression of pack ice and/or longer open water periods continues." This observation was yet another warning flag. According to Amstrup, it provides "concrete evidence of what we knew before, that there are limits to the amount of water exposure polar bears can take, and that they can indeed drown. . . . Polar bears are good swimmers and can swim long distances under the right circumstances. Long swims for them must be unusual and stressful, however. As they are not aquatic mammals like seals and whales, they are not nearly as well equipped for extended durations in the water. . . . The situation now, how-ever, is that there is far more open water in their environment than historically (at least for the last 5 thousand years or so)."[10]

In his latest report on bears along the Alaskan and Canadian coast of the Beaufort Sea for the U.S. Geologic Survey, Amstrup, along with Stirling and Eric Regehr, says that fewer cubs are surviving now. The percentage of very young cubs found alive after a year has dropped from 65 percent to 43 percent.

Mounting research shows that the world's 22,000 polar bears, the largest land carnivores, are being affected by the diminishing polar ice cover. Scientists in northern Europe and eastern Canada have documented that bears are forced by melting sea ice to move to land earlier and stay there longer and thus do not find enough food to build fat reserves to survive. The bear is a candidate for "threatened" species in the United States. (AUGUST 2002)

Moreover, comparison with a similar sample of bears captured in surveys made before 1990 shows that cubs and adults weigh less now. This is ominous, because the same changes were noted in the Hudson's Bay population long before its bears started to decline. At the same time, the risk of conflict between humans and bears will likely grow. In recent summers, three quarters of Alaskan bears sighted were fewer than 8 miles (12 km) from Kaktovik, a Native village on the coast, within the Arctic National Wildlife Refuge (ANWR). A third of Alaskan polar bear dens were found in the part of the refuge that may be thrown open for oil drilling. According to Scott Schliebe of the U.S. Fish and Wildlife Service (USFWS), more bears are now seen on land compared to the period 1970–90; the number, moreover, correlates to the distance between shore and pack ice.[11]

Long-term studies of polar bears to the south along Hudson's Bay shed troubling light on this situation. The animals there are now being forced by fast-melting sea ice in summer to move onto land more than two weeks earlier than thirty years ago, without having developed healthy fat reserves. And the fall freeze-up is now occurring nearly a month later. Because most bears fast while on land waiting for the sea to refreeze, at the end of summer the now very skinny animals are in poor condition to survive the winter. Not only that, but underweight females produce cubs with low birth weight and, hence, reduced survival chances. In a 2004 paper summarizing research done near Hudson's Bay, Andrew Derocher of the University of Alberta and coauthors said: "Given the rapid pace of ecological change in the Arctic, the long generation time, and the highly specialized nature of polar bears, it is unlikely that polar bears will survive as a species if the sea ice disappears completely as has been predicted by some."[12]

Reports such as this, along with the continuing shrinkage of summer Arctic sea ice, have put the polar bear on the U.S. threatened species list under the Endangered Species Act. The USFWS made the decision in May 2008 after a petition and several lawsuits by the Center for Biological Diversity and other NGOs. The government agency did not cite global warming as a proximate cause, and it clearly signaled that it would not apply the Endangered Species Act to greenhouse gas emissions or "abuse" it to make global warming policies.[13]

THE GREAT HERDS OF CARIBOU still follow the wilderness migration tracks they established as the glaciers receded eighteen thousand years ago. They are a remnant of the Pleistocene fauna, which once stretched from North America through northern Europe and into China. The saber-toothed tigers and mammoths are gone, but, along with the great bears and musk ox, the caribou remain. Caribou, or reindeer, are found in tundra regions of North America, Siberia, Greenland, and Scandinavia. Their range also extends into the boreal forest far south of the Arctic Circle. In North America, thirteen herds totalling three to four million animals wander the tundra, giving birth at specific times and locations—a trait that makes them susceptible to climate changes. Perhaps the most famous is the Porcupine herd of northeastern Alaska and western Yukon. Each year, this group undertakes the longest migration of any mammal, more than 2,700 miles (4350 km) in a circuit across the Brooks Range. Although the population of the Porcupine herd has varied, since 1989 it has fallen from almost 180,000 animals to about 120,000 now. The role played by changing climate in this decline is unclear. Earlier snowmelt, leading to earlier spring plant growth, appears to be a positive factor for caribou. Because cows require twice their normal

Caribou wander past a flock of black brant who are waiting out their molt in wetlands near Teshekpuk Lake, Alaska. This area east of Barrow near the Colville River delta is one of the most important bird nesting areas in the world. It is imperiled not only by invading salt water and shore erosion but also by proposed oil drilling. [AUGUST 1997]

energy input during peak lactation, the availability of food increases their ability to feed their calves, so the survival rate goes up. Recently, in step with earlier greening at the birthing area in the Arctic National Wildlife Refuge, plant biomass has been up some 50 percent in late June. However, plants that green earlier tend to die back sooner, which may be reducing caribou nutrition going into the winter.[14]

Winter is becoming increasingly difficult for caribou. To the east of the Arctic Refuge, a herd of smaller Peary caribou in northern Canada has been hard hit by deeper snow on winter feeding grounds, a consequence of warmer temperatures. Observers of the Porcupine herd have documented, too, that warm spells result in layers of ice that make it harder for the animals to reach forage under the snow and to escape predators. In several recent years the females have been delayed by heavy late-winter snows and forced to calve away from their preferred location on the northwest coastal plain of the Arctic Refuge. Even though this has resulted in fewer surviving young, the question of whether unusually severe winter events are affecting the Porcupine caribou remains unanswered, said USGS caribou scientist Brad Griffith; there appear to be "delayed and complex interactions between climate, caribou habitat, and population performance."[15]

Scientists are just beginning to see the effects of climate change on other Arctic wildlife. The ten million geese and many millions of shorebirds and ducks that migrate into the tundra each summer to nest are of international importance. Most of them migrate from temperate-zone nations as far away as Argentina and New Zealand. So far regional warming and lengthening of the summer season have proved advantageous to birds as varied as the tundra swan and various flycatchers. The World Conservation Monitoring Center warned in 2000 that tundra nesting areas for endangered birds such as the red-breasted goose and spoon-billed sandpiper will shrink in size by up to 85 percent if the climate changes as predicted—because of sea level rise and coastal erosion in the north, as well as forest encroachment from the south. In Alaska, interest and funding for research tends to focus on proposed oil-drilling areas. One team of scientists has been going to the Teshekpuk Lake Special Area in the U.S. National Petroleum Reserve near Barrow since 1979, not only to repeat inventories of birds but also to survey lakes and ocean shorelines. This is very wet tundra with meandering streams and large lakes on which 100,000 brant and other geese take refuge during their vulnerable flightless molt. "Dramatic amounts of erosion have occurred in some places," the 2005 progress report states. "For instance, at the J. W. Dalton test well in north central [an area in the survey], over 60 meters of coastline were lost just between 2003 and 2004. . . . The amount lost since 1979 is far greater. Many other examples of coastal shoreline changes have been documented. . . . We hypothesize that increased rates of coast erosion are affected by reduced sea ice extent, as ice forms a buffer from waves, and from increasing thawing of permafrost which makes the land more vulnerable to wave action." Geese throughout the Alaska coastal region feed on eelgrass (Zostera marina) and algae that can be disrupted by this tidal and storm damage. Ocean erosion led to the saltwater flooding of two freshwater lakes, an "intrusion [that] is expected to quickly alter foraging habitats for geese," because around Teshekpuk Lake birds rely heavily on freshwater sedge and grass during their molt. Goose populations are relatively stable so far, said the report, but the birds "have varied in number and redistributed themselves, potentially in response to the types of physical and plant community changes we detected."[16]

CONCERN ABOUT SPECIES SURVIVAL often focuses on ecosystems characterized by high diversity such as occur in the tropics, but the Arctic, with its less complex natural communities, is equally at risk. Arctic plants and animals depend intimately on one another, so when one species is lost from the system, others may soon follow. Also, global warming will likely drive competing animals, plants, and disease up from the south. A simple change—in temperature or the time of flowering, for example— can have a disproportionate impact and may well lead to profound consequences throughout the ecosystem. One dramatic example involves the great Chinook salmon runs of the Yukon River. Up to a third of adult Chinook salmon entering the Yukon River in the years 1999 to 2002 were found to be infected with *Ichthyophonus*, a deadly parasite. "Prior to this," study leader Richard Kocan of the University of Washington told a science forum in 2003, "it has never been recorded in salmon anywhere except by artificial transmission." Visible signs of disease were minimal when fish entered the river but increased markedly when they reached Rampart Rapids at river-mile 745. Many of the diseased fish did not live long enough to spawn. Peak temperatures above 60°F (15°C), which is very warm for salmon, were frequently observed in this stretch of the river. "Examination of historic temperature data," the final report states, "suggests that rising average [Yukon River] water temperatures during the past three decades appear to be associated with the increase in disease."[17]

Out in the ocean, where the salmon live up to eight years before returning to their spawning grounds in the Yukon, water is also heating up; it is more than 3.5°F (2°C) warmer than it was only ten years ago. Winter ice now melts three weeks earlier. "The Bering Sea," researchers reported in 2004, "is experiencing a northward biogeographical shift in response." Since then, other scientists saw the northern Bering's icy, shallow-water ecosystem, favored by eider ducks and walrus, becoming more open and inhabited by fish from the warmer south. Rich sea-floor ecosystems face nutrient changes. Gray whales are shifting their feeding patterns northward. Walrus, which prefer to stay among ice floes, can find the ice drifting far from usual feeding areas. These changes, of course, affect the native subsistence hunters of the region. Although some researchers suggest (as for Arctic Ocean sea ice as well) that the warming began in a cyclical phenomenon over the Arctic Oscillation, what may have begun as a natural shift might not recover well in the face of persistently rising temperatures from global warming. Feedback of heat from open water and snow-free land is also strengthening. The consequences could be serious for the Bering Sea, which accounts for almost half of U.S. fishery production (by weight) and is habitat for most of its larger sea mammals and 80 percent of American seabirds, from albatrosses to storm petrels.[18]

BOREAL FORESTS IN FLAMES

Other ecosystem changes, some of them running counter to previous assumptions, are rampant in the forest zone just south of the tundra. The boreal forest is equal to the tropical forest zone in its influence on the Earth's oxygen supply and water cycle, though it is not quite as large. Covering more than 6 million square miles (15.5 million sq km), it forms a band around the northern continents just below the Arctic Circle, dipping as far south as 45°N in Siberia, where it is called taiga. A third of the boreal forest is in Canada and Alaska. This stretch of spruce, larch, fir, and pine, along with alder, birch, and

Nearly four million acres of mature white spruce forest on the Kenai Peninsula in Alaska have been killed by a growing population of spruce bark beetles *(Dendroctonus rufipennis)* since about 1987. Scientists attribute the native beetle infestation to rising average temperatures in both winter and summer. (JULY 1999)

One result of beetle attacks and higher temperatures has been increasing spread of wildfires. This fire near Tok was part of the largest acreage ever burned in Alaska in a single year. Forest death and burning is increasing in British Columbia and parts of the American West—in fact, throughout the world. Including grassland fires, a record 9 million acres burned in the United States in 2006. (PHOTOGRAPH COURTESY ALASKA DIVISION OF FORESTRY, TOK OFFICE)

aspen, is of crucial importance to many mammals and migratory songbirds. Its main feature in a climate narrative, however, is that even though it covers only about 15 percent of the Earth's land surface, it stores one-third of all land carbon—690 billion tons (625 billion mt). Its trees and plants take up great quantities of CO_2, release the oxygen, and store the carbon in their cells (wood is half carbon) and in the duff and dirt of deep, cold soils. Decomposition of this material, which recombines the carbon into CO_2 and methane, is so slow in the north that very little of this carbon gets out again.

According to a report issued in 2003 by the United Nations, however, "the degradation of permafrost is widespread" in boreal zones in China, Mongolia, and North America. "Natural stands of paper birch on these soils die and aquatic species invade, forming lowland fens and bog meadows within 30 to 40 years." The new bogs store carbon in accumulating organic matter and begin to release methane because of wet decomposition conditions, as documented in Alaska, Sweden, and the Russian tundra. Another visible result of climate warming is the migration of this forest northward (and up in elevation)—a vast advance of trees into tundra landscapes that is being documented across the Arctic. Satellite studies show that about 15 percent of the tundra—an area three times the size of California—has been lost to the advancing treeline since the 1970s.[19]

The greatest threats to the boreal forest are fire and insect attack. The average area destroyed by fires in the North American boreal forest region has increased from 3.7 million acres (1.5 million Ha) a year in the 1960s to 7.5 million acres (3 million Ha) in the 1990s. Record heat in the early 2000s has continued the trend. Summer 2004 was both the warmest summer on record in Alaska and the worst by far for wildfires in the state. According to University of Alaska botanist Glenn Patrick Juday, 6 million acres (2.7 million Ha) burned—more than 10 percent of the forest cover in Alaska's interior—and 8 percent of the forests in the Yukon burned. The fires he saw burned "from horizon to horizon; they burned wetlands, krummholz (alpine trees), everything. The fires completely washed over the landscape." Huge burns swept across the vast boreal taiga of Siberia as well: University of Maryland geographer Eric Kasischke estimated that 54 million acres (22 million Ha) were burned in 2003 and 40 million acres (16 million Ha) in 2005. Lowering water tables and drying peat exacerbate the burning, as fires eat down through the vegetation understory and into the soil, where they may smolder for weeks in the rich organic matter. The billowing plume of smoke contains a huge amount of carbon dioxide (and carbon monoxide, which converts to CO_2), up to 150 tons per hectare. Thus, in a year in which there are large fires in the boreal region, well over a billion tons of CO_2 can enter the atmosphere. Afterward, the blackened stands soak up solar radiation, which melts the permafrost, which leads to even greater carbon dioxide and methane emissions. Thus, as frequency and size of natural fires in the boreal forest increase in step with temperature, a feedback loop proceeds to enrich the air with even more greenhouse gas. That vicious cycle has also released another evil into the pall of smoke from boreal fires. Research reveals that these forests have been dusted by mercury emissions throughout the industrial age and that the mercury has settled into the cold soils, where it remains inert and out of the food chain. But with increasing dryness and fires, up to fifteen times more mercury is being released back into the air now than previously measured over the north.[20]

Fires could envelop nearly 4 million acres (1.6 million Ha) of mature white and Lutz spruce forest on Alaska's Kenai Peninsula and throughout south-central Alaska, trees that have been killed since 1987 by a growing population of spruce bark beetles *(Dendroctonus rufipennis)*. The U.S. Forest Service says it's the worst beetle outbreak in Alaskan history, affecting logging, tourism, and property owners. Already this very active event has spread to more than thirty-eight million mature spruce trees in Alaska, and has reached the Yukon and other parts of western Canada. The beetles' larvae devour the rich cambium layer, girdling the trees and turning entire mountainsides from verdant green to dull red, then gray as needles fall off. The insect is native to these forests and usually lives in balance with the trees; although tree rings give evidence of occasional mass tree kills in the past, there has never been anything like this one. Scientists Edward Berg, an ecologist with the Kenai National Wildlife Refuge, and Wisconsin entomologist Kenneth Raffa correlate the infestation with rising average temperatures, in both winter and summer. According to Berg, "The warm summers of the 1990s are the longest run of warm summers in the last 350 years." Summer temperatures at the Kenai airport are up almost 3°F (1.7°C) since the 1940s—but winter readings are as much as 9°F (5°C) higher. This means more beetle larvae can survive to spring, when there are plenty of warm days for the beetles' mating flights and egg laying; the high summer temperatures then allow them to mature faster and complete a two-year life cycle in one year. The trees, in contrast, are weakened by the hot, dry weather and do not have enough natural defenses to withstand the beetles' assault.[21]

Fire and bugs are not the only threats to the boreal forest, unfortunately. All over the Northern Hemisphere tree ring densities declined in the second half of the twentieth century, indicating that something is limiting growth. One study focused on Alaskan spruce, which has been under stress from "severe and prolonged warmth and dryness from the 1970s to the present that is unprecedented in the twentieth century," as noted by University of Alaska researchers Valerie Barber, Glenn Patrick Juday, and Bruce Finney. This is causing even the fastest-growing white spruce trees to put on thinner annual rings. "White spruce is one of the most productive and widespread forest types in the boreal forest of western North America," the scientists wrote in *Nature*. "Thus any coherent climate-related change in white spruce growth is likely to be an important factor in CO_2 uptake in the boreal forest, a region that is one of the planet's major potential carbon sinks." Juday said starkly: "Trees are dying because it's too dry and too hot for them." What these trends predict "is the death of the trees within the century."[22]

AS THE ICE DISINTEGRATES, SO DO ARCTIC WAYS OF LIVING

Extension of warm weather and changes to ice and wildlife affect Natives, residents, and workers in the north. Hundreds of buildings, roads, and telecommunication towers in Alaska and Siberia are tilting and deforming as the once-solid permafrost weakens. In Yakutsk—where warming is estimated at about a quarter of a degree Celsius per year—effects include severe damage to many multistory buildings built on the ground. In Fairbanks, roads and bike paths are cracking and settling into roller-coaster rides. On some streets, including the aptly named Madcap Lane near the University of Alaska, every house has been torqued or had part of its foundation cave in as the permafrost thaws underneath. Geophysicist Tom Osterkamp, however, warns that we should not make too much of these collapses, because the warmth from the structures themselves can cause the permafrost to thaw. Where permafrost is warming, such as along Alaska's North Slope, his studies show no overall increase in the amount of active-layer melt-out (referring to the surface layer that thaws each summer). For some residents of the north, such as farmers in Greenland, the warming has been a boon, increasing farm areas and lengthening the growing season.

Local conditions vary a great deal, though, and warming permafrost in concert with earlier spring thaw is creating problems for many in the Arctic. The oil industry has been forced to take note, because most oil exploration and some drilling must be completed in the coldest part of winter, when the hard frozen tundra resists most damage to the terrain from heavy equipment and when ice roads can be made by spraying water over the ground. This window for work is getting shorter. One official response was an effort by the state of Alaska and the federal Department of Energy in 2003–4 to extend the work period by revising frozen-soil standards and using reduced-impact oil-drilling vehicles. "During the past three decades," read the project justification, "the number of days between the opening and closing of the tundra for exploration activity has declined from over two hundred days in 1970, to only one hundred three days in 2002. . . . This trend appears consistent with findings of general warming in the Alaska arctic associated with global climate change."[23]

The increasingly open water of the Arctic Ocean is likely to upset the political and economic balances of the north. It will, for example, become much easier to extract minerals and oil from coastal areas as they become more accessible to large ships. Unknown islands hidden beneath the ice are being discovered by ships and satellite examination of the thinning sea ice. North of Scandinavia and Russia, the sea is already opening up for summertime shipping, offering a route to Asia that is 40 percent shorter than the course through the Suez Canal. The fabled Northwest Passage, which fueled so much exploration of North America, is also predicted to open yearly, shortening the journey from Europe to the Pacific by some 6,800 miles (11,000 km) compared to the Panama Canal route. Already a great many ships are plying the Arctic. Since 1977, according to researchers for the Arctic Council, sixty ships have reached the North Pole. And traffic through the region has been steadily mounting. In 2004 alone, 11 oil-exploratory icebreakers, 165 commercial ships, 27 cruise ships, and 6 scientific-research icebreakers plied Arctic waters. This burgeoning traffic, the discovery of new islands, and the hunt for mineral deposits will surely test the sovereignty of Arctic nations. Canada is responding to this possible encroachment by mounting a series of military exercises in its northern provinces. One large deployment brought a naval frigate, squadrons of aircraft, and more than 400 personnel to Baffin Island in August 2004, and other exercises followed even farther north. In a speech at Station Alert, the northernmost permanent settlement in the Arctic, Canadian Prime Minister Stephen Harper declared that "sovereignty is not a theoretical concept; you either use it or lose it."[24]

The 400,000 Natives of the north are experiencing the thaw in all facets of their lives. As documented in the Arctic Climate Impact Assessment, even though their ancestors lived here at least twenty thousand years ago and over time withstood many climate changes, the current warming is so rapid it strains human resilience. The very ground and waters these people live on and hunt over no longer conform to their ancient knowledge of seasons and weather. The loss of strong ice at the beginning and

Kivalina, an isle-bound village of Iñupiaq on the Chukchi Sea in western Alaska, is eroding away like Shishmaref to the south. Natives have lived on this sandy barrier island on the northwest shore of Seward Peninsula for many hundreds of years, hunting seals from camps that became villages only in the mid-twentieth century. [AUGUST 2005]

end of winter threatens their ability to hunt and engage in subsistence food gathering. More and more, hunters and travelers fall through ice that they once knew as secure. The wildlife they hunt is changing its habits due to shifting forage and prey availability, and other food such as fish and berries are often of lower quality now. In Alaska, residents of 184 out of 213 Native villages are facing the possibility that the land on which their homes stand will erode away or be flooded out as sea level rises, permafrost thaws, and rivers change character due to rising temperatures. According to a 2003 Government Accounting Office study, the dangers are most vivid in Kivalina, Koyukuk, Newtok, and Shishmaref, all of which face relocation. Five other towns, including Barrow, Bethel, and Kaktovik, where average temperatures in the last thirty years have increased 3°–4°F (1.7°–2.5°C), are having to cope with increasing damage. The greatest problem is loss of shore ice, which allows storm waves and ocean swells to pound the sandy shoreline.

Erosion from effects of warming is most acute at Shishmaref, a hamlet of 590 Iñupiaq on the edge of the Seward Peninsula facing the Chukchi Sea. The village is perched on a sand spit. For hundreds of years the people came in spring to hunt seals off the sea ice as it began to melt. The animals were flensed with the Native ulu knife, and the meat, intestines, and skins were dried on wooden racks. Each family had its own spot along the cliffs. When the State of Alaska required formal education for all citizens and built a school in the mid–twentieth century, a town of small houses coalesced around the new building. The traditional sealing and other subsistence activities continued into today's world of modern jobs, four-wheelers, regular air service, and electronics. Now the permafrost beneath the dunes on which the village is built is thawing, and rising seas and more ice-free water invite stronger waves to eat away at the quarter-mile-wide (400 m) spit of sand during storms. Erosion has been severe since the 1950s; cement armor and wire gabions installed by the U.S. Army Corps of Engineers have proved ineffective. Luci Eningowuk, chair of the Shishmaref Erosion Relocation Committee, said they often lose 60 feet (18 m) of shoreline a year. The town sewage pond, only 20 feet (6 m) from the shore, is seeping to the sea. In a single storm in 1997, 125 feet (37 m) of beach and dune washed away. Townspeople struggle to move their houses away from the shoreline, but storms keep eating away more and more sand.

I first visited Shishmaref in 2001 to document this erosion and found the beach littered with twisted concrete and the wire remains of failed gabions. A raw fresh sand cliff sliced across the west end of town, right through several buildings; utility pipes and cables dangled in the scarp. When I revisited the village four years later, almost nothing was left of the west end. A single small house hung over the cliff edge, ready to topple the 12 feet (3.5 m) to the beach. None of the houses that had topped the cliff in 2001 remained; all had been moved or had fallen victim to large waves beginning in 2002.[25]

To get a sense of what has been lost, I climbed into Tony Weyionanna's hand-built boat on the bay side of the village. He motored around the spit into the open Bering Sea, bore south, and stopped off shore of the village. We gazed several hundred feet across the water at the remaining sandy bluff with its scattering of buildings. "Thirty years ago," he said, "we played ball right over there," and he gestured toward the water near the boat. Tony's face reflected not nostalgia, however, but his concern for his fellow villagers, who in 2002 voted to move to more protected ground away from the ocean—and away from their traditional home, fishing, and sealing sites. "Winter comes two months late; spring a month

The Iñupiaq who inhabit the village of Shishmaref, Alaska, have survived here for generations but can't halt the rising Bering Sea and the thawing of permafrost within the dunes. In 2002 residents voted to move their village to higher, more protected ground away from the ocean, giving up their traditional home, fishing, and sealing sites. (JUNE 2001)

earlier," Tony said, underscoring the changing seasons that have upset the food-gathering cycle of this village. He is among the residents of Shishmaref who will be responsible for carrying out the relocation. Studies are continuing about how best to make the move, which will cost $150–$180 million—a sum that Alaskan and federal agencies are still trying to raise. No one has totaled up the funds needed to adapt or move the other hundreds of Native villages and thousands of indigenous people whose homes in the United States, across Canada, and into northern Europe and Siberia are becoming ever more precarious.[26]

"THE LAND THAT IS MELTING"

On Baffin Island in the far northeast of Canada, in Nunavut, another village finds itself on the edge of change. Pangnirtung, an Inuit hamlet on a narrow fjord surrounded by glacier-carved cliffs, has seen wrenching cultural shifts before. In August 2004 I spent a week there trying to understand the challenges this community has faced in the past and is now facing from climate change.

The elders of this village of 1,300 have sharp memories and detailed knowledge of their ancestral homeland. Elisapee Ishulutaq, a spry seventy-eight-year-old artist with a radiant smile, was born in an outpost camp when most Natives here were nomadic. She recalls when her family would come to Pangnirtung to trade seal skins and eider eggs for tools, ammunition, and fuel at the Hudson's Bay Company. They were dogsledding as late as July, more than two months after the ice breaks up now. "All the mountains were covered with glaciers," she said. "There isn't any deep snow anymore."

Previous changes in the area—the shift from nomadic hunting to whaling, for example, or the introduction of twentieth-century Western urban culture—left intact the relationship between the Inuit and nature. People could count on natural events: the long winter that froze the sea and bays hard and early and did not release them until June; the solidity of the land that, underlain by permafrost, provided secure home sites; the ever-present snow and ice on the mountains through summer, which regulated the flow of rivers; the cycles of plants and animals. All these natural cycles are now being distorted. For the 155,000 Inuit and Iñupiaq people across the Arctic, from Siberia to Alaska, the changes mean probable loss of their hunting culture, insecurity as to food and shelter, and devastating disruption to health and social structures.

Change came initially to Pangnirtung, according to official village history, when Scottish whalers sailed into the waters of Baffin Island in 1838. The Inuit's nomadic life of thousands of years quickly gave way to hunting bowheads. Less than one hundred years later a Hudson's Bay Company trading post and a hospital encouraged Inuit to migrate from traditional hunting camps. The Canadian government established a school. Many families resisted settling in town until, in the 1950s and 1960s, distemper forced the killing of many of their teams of essential sled dogs. Some Inuit think the Royal Canadian Mounted Police shot many more dogs than necessary in order to push them into the towns (an idea discounted by a 2005 government investigation). In any event, the dog slaughter marked a wrenching transition for the hunters and their families, severing a deep connection to their historic way of life. Since then the shift into the technical and electronic world culture of the twenty-first century has been rapid, if incomplete.

Pointing to photos of herself as a young mother, Elisapee Ishulutaq recalls summers in Pangnirtung that did not begin until June and days throughout the warmer months when everyone had to stay bundled in sealskin clothes. Now midyear days are frequently warm enough for light coats, and Elisapee says the snows are gone from the surrounding mountains. (AUGUST 2004)

Pangnirtung sits on a rocky point astride a long fjord carved ten thousand years ago by a lobe of the Laurentide Ice Sheet. There are no paved streets, and not even the airport enjoys the benefit of asphalt. Water is brought to each house by tanker truck, and a different set of trucks pump out and take away sewage. There are three stores in town, which sell most of what you'd expect to find at a K-Mart, from frozen food to fishing gear, but in smaller quantities and at higher prices. Most families enjoy television, computer access, and a lot of outdoor technology like high-powered rifles, GPS units, and all-terrain vehicles. Many large skiffs are anchored in the harbor and perhaps even more boats are pulled up into backyards, most with large outboard motors. These are frequently parked next to snowmobiles, together constituting tools of the modern hunter—and connecting the millennia of nomadic survival in this Arctic landscape with today's world. Unlike millions of Americans and Canadians far to the south who have boats, snowmobiles, and GPS for recreation, the Inuit still survive by hunting and fishing.

Pangnirtung's location gives it a grandeur—and perhaps a source of hope—that most other villages do not share. This is the gateway to Auyuittuq National Park, 35 miles (55 km) to the north, a

7,300-square-mile (19,000 sq km) temple of the last ice age, a Yosemite only partly melted out, accessible only by boat or snow machine. Tourism to the park is an increasing source of income for the hamlet, and a number of local hunters have recently become certified guides and outfitters. Still, at present no more than four hundred visitors a year come, according to park interpretive ranger Billy Elooangat, no doubt because of the park's remoteness and the great cost of the trip.

Yet here is another area where the north, its myth and its landscape, is feeling the hot breath of global warming. The glaciers grinding down from the Penny Ice Cap into the park valley are shrinking rapidly. Ranger Elooangat cited geologists' measurements showing that Turner Glacier, above Summit Lake, receded 25 feet (8 m) just between 2002 and 2003. Awareness of glacial melting, he said, is "crucial for safety reasons. A faster rate of melting of the glaciers creates big rivers; that's our main concern. With a little bit of rain this creates a problem for hikers." He traced the trail that passes under four glacier tongues on the way into the park, then looked up from the map. "*Auyuittuq* means 'the land that never melts,'" he commented, "but we are going to have to call it 'the land that is melting.'"

AT PANGNIRTUNG, elders like Elisapee Ishulutaq and other villagers I spoke with said that they would use the past tense: the ice is nearly gone for them. Only tiny patches of ice remain visible from town in summer. In winter, the ice the Inuit need to cross the fjord and big rivers is thick and strong for only a few months now, from December to May rather than November to July. The effects are wide ranging.

AN EVER-WARMING ARCTIC?

JONATHAN OVERPECK

Institute for the Study of Planet Earth

The last time I was on Baffin Island, over a decade ago, I awakened one morning in a remote tent camp near the Arctic Circle to the sound of sea birds and running water. Never before on my springtime research expeditions had I heard such sounds. Baffin Island should have been frozen solid this time of year, providing ideal conditions for our sampling of lake sediment from ice and for getting around on snowmobiles.

In the days that followed, our research area proceeded to melt faster than I could ever have imagined. Although previously Baffin had been slow to show signs of warming, something had changed. By 2006,

my earlier musing proved right: the entire Southern Greenland, Baffin Island, and Labrador area was warming dramatically.

The evidence was alarming. Retreat of Arctic summer sea ice had started to accelerate. Declassification of Cold War–era submarine data revealed that some sea ice had thinned by a shocking 40 percent over the last fifty years. Permafrost was melting in more and more locations. Arctic glaciers and small ice caps were in noticeable retreat. And, most interesting, the Greenland Ice Sheet was melting and losing mass at an increasing rate, especially in its southern reaches. The Arctic is clearly melting, and it is doing so ever faster.

My surprise of a decade ago changed to a drive to figure out what all the recent Arctic change means for the future. I joined a large group of colleagues to examine how

recent Arctic warming fit the pattern of global temperature change going back centuries. Our first comprehensive analysis revealed that late-twentieth-century Arctic warming was unprecedented over at least four hundred years. Not only that, but it was part of an unprecedented warming of the entire globe. No natural cause has been found to explain this warming—the Arctic is melting because of human emissions of greenhouse gases.

Thanks to researchers at NASA, we now know that 2005 was the warmest year on record in the Arctic. In the last fifty years alone, the average temperature in the region has risen more than 3.5°F (2°C). At this rate, warming over the next one hundred years won't be minor, or even moderate. Humans could drive the Arctic to be warmer than it has been in at least a million years.

Seventy-five-year-old Jamese Mike is a retired hunter who has kept daily weather records for years. Sitting at his kitchen window with binoculars ready to check events out on the fjord, he fanned out a stack of pocket notebooks thick with penciled information. "The glaciers that were on the mountain-top ever since I can remember are gone," Jamese said. "I notice the winds are shifting around a lot and it is raining a lot. This is not what we expected. The rain is causing more rivers to flow."

Peteroosie Qappik was born seventy-two years ago in Nunataaq, a nomadic site to the west of Pang-nirtung. He was always a hunter, with no other employment. "I remember the weather used to be nicer and more beautiful," he said, "and in August, the most beautiful. It was almost never windy." Now, however, "the wind has shifted. The strongest used to be from the north, but now it's from the north-east." Sitting in the museum of the community visitor center, Peteroosie gazed out at the fjord, which freezes over so much later now. "The ice conditions are different. The changes are coming from the bottom. The breathing holes for the seals are pockmarked, because [the ice] doesn't get as hard." He went on, "I'm not so sure about the seals. Because of the ice changes, the baby seals are not even grown when the ice breaks up in May. Long time ago the ice didn't break up until July, and it was a lot thicker. Now it's only half as thick because there isn't as much time to freeze. And it melts from the bottom. The older people know for sure the water is warmer."

These observations were repeated by Jaco Ishlulutaq, a fifty-six-year-old hunter whose family brought him to Pangnirtung from a nomadic camp when he was a young boy. "When the ice doesn't

With this magnitude of warming, the new Arctic will no longer be a place defined by year-round sea ice, extensive permafrost, many glaciers and ice caps, or even, conceivably, a Greenland Ice Sheet. Recent research suggests that current trends in the burning of fossil fuels will, sometime later in this century, push the Earth across a threshold beyond which irreversible melting of the Greenland Ice Sheet will likely occur. This melting will take centuries or longer, and it will cause global sea level to rise by up to 23 feet (7 m) above present levels. Other scientists have shown that it might take *hundreds of thousands* of years to get the natural Arctic—the Arctic we inherited—back.

Such environmental change will devastate many Arctic plant and animal species. The polar bear and many less charismatic species are poorly equipped to survive in a warmer, iceless world. Unprecedented ecological change has already been documented in the oceans and lakes of the Arctic, especially in coastal environments, and across the tundra into subarctic forests.

Among the biggest losers at the hands of future climate change will be the native human inhabitants of the Arctic, the Inuit. Already, climate and associated sea level change are wreaking havoc on the Arctic's native peoples, and they have been among the most vocal in calling for action to stop global warming. Climate and sea-ice change will increasingly cause near-shore declines in animal populations—seals, walruses, polar bears, and birds—on which the Inuit culture depends. Subsistence food sources on land, including select plants and animals, are at similar risk as permafrost melts and seasonal cycles shift. No wonder some Inuit see the fight against global warming as a fight to save their very culture.

There is no longer any question about cause and effect in all this: the change in the Arctic stems from humans polluting the atmosphere with ever-increasing amounts of carbon dioxide and other greenhouse gases. As disturbing as this picture of Arctic ecosystem change is, we do have choices. Through advanced technology and energy conservation, developed countries like the United States have the resources to lead and act in ways that will keep the changes I've described, and many others, from being our legacy to future generations. What happens next lies in our hands. ✦

form," he said, "the animals that live on it will have to change. Their fur will have to change. The plankton that live under the ice will disappear, and that changes the food chain." Jaco said most hunters notice that plankton—*ijiitoo,* meaning "little things with big eyes"—are decreasing. "We have to respect plankton, and respect the ice, because the ice is their homeland. Ice is where they breed, live, and survive." Jaco also said flowers are coming out sooner. "The arctic willow is growing longer, with larger leaves, and taller. Up to a foot and a half."

Not only the elders notice. On a building site I met Geeold Kakkik, who was running a pile driver for metal pipes that would be sunk into the permafrost to serve as the building's foundation. Structures built on the ground, even on foundation blocks, shift as the ground freezes and thaws, so most construction now rests on these five-inch pipes. However, the depth in the soil at which permafrost begins is getting greater with each year, Geeold said. "Last year in July the pipe went down four to five feet to frozen ground. Right now it's six or seven feet, and some spots are deeper than that."

This deeper active layer of soil is a physical danger because it is beginning to slump and cause landslides just above town. Community lands administrator Bill Kilabuk oversees planning functions and is involved in utilities and power generation. He confirmed that the permafrost is receding. In the future, he said, erosion will be a great problem. "In the last several years, there has been more erosion in some parts of town. Global warming melted the permafrost and allowed the land to loosen and slump down." Taking a more positive stance, Bill added, "The snow, the seasons are still here. They just happen later."

Walking through town, I could see the effects of slumping. I was also blasted by sand whipped up by wind that the elders said was much stronger than in their youth. In historic photos of Pangnirtung, kids are seen dressed in warm furs even in summer as they played Native games. Today, however, in a strongly symbolic, incongruous vision of change here in the north, I watched youngsters out on a shoreline meadow in their hooded sweat suits and running shoes, playing makeshift golf.

The more intense, sinister changes—loss of the old hunting and sled-dog culture and the resulting social and family pressures—may be exacerbated by climate shifts in Pangnirtung. Jeanne Mike, Jamese Mike's daughter and a social worker, thinks there "definitely is" a correlation between domestic abuse and climate changes, especially during the fall, a time when ice used to form and men headed out with their dog teams to hunt. "The guys are stuck in town. They can't go out. The suicide rate is up, too," she said. "Peak suicide rate is [during] spring and fall." Pangnirtung, small though it is, has several suicides a year, with a rate ten times the Canadian average. Most of the victims are men aged fifteen to twenty-four. Root causes, according to studies, include lack of education and high unemployment. Climate change, making it ever more difficult to hunt subsistence animals, could severely aggravate these factors.

Jaco Ishlulutaq's final words to me reflected the many changes the Inuit have survived, as well as the native optimism of their culture. "There's a myth, an old saying, that the planet will shift and change. But we should not be pessimistic about it, but keep going. We should not be angry at changes, because the planet gives us what we need. Go with it."

It's true that changes are nothing new to the Inuit, Jamal Shirley of the Nunavut Research Insti-

tute in Iqaluit told me in a phone conversation. "The people of the north have a history of adapting to change, sometimes catastrophic change," he said. "Dealing with change and unpredictability, times of scarcity, has been a way of life for many years." Nevertheless, people in many places are reporting climate shifts in the last fifteen to twenty years especially, and they are saying that "changes have been more visible, more evident, more rapid." As one elder put it, "This has really affected the hunter's way of life."[27]

Sheila Watt-Cloutier, former chair of the Inuit Circumpolar Conference, sees incalculable harm from changes that may completely disrupt the society and subsistence life of the north. During her term of office, from 2002 to 2006, speaking for the world's 155,000 Inuit, she lobbied governments and international leaders to control global warming emissions. In December 2005, fed up with the obstructions and slow pace of action, she and sixty-two Inuit elders and hunters appealed to the Organization of American States Inter-American Commission on Human Rights, seeking relief for all Inuit from violations of their rights resulting from global warming caused by the United States. Their petition, a 175-page document with 796 scientific and legal footnotes, lays out the facts of Arctic climate change as stated in the Arctic Climate Impact Assessment and other studies, as well as generally accepted statistics regarding the United States' role as the leading greenhouse gas producer. "The impacts of climate change, caused by acts and omissions by the United States, violate the Inuit's fundamental human rights protected by the American Declaration of the Rights and Duties of Man and other international instruments," it reads. "These include their rights to the benefits of culture, to property, to the preservation of health, life, physical integrity, security, and a means of subsistence, and to residence, movement, and inviolability of the home." Watt-Cloutier, who was born in Kuujjuaq, a village in northern Quebec, called the petition "perhaps the most loving act I have ever brought forth," because so much is at stake—namely, that "my grandson may not have what I had, that gave me the culture and strength to do what I do." In announcing the appeal at the 2005 climate change conference in Montreal, she said it is about the Inuit, "a powerful, resilient, resourceful and wise culture," not "becoming a footnote to globalization." The preceding year, she testified before U.S. Senator John McCain and the Senate Committee on Indian Affairs, saying that "climate change in the Arctic is not just an environmental issue with unwelcome economic consequences. It is a matter of livelihood, food, and individual and cultural survival. It is a human issue. The Arctic is not a wilderness or a frontier. It is our home and homeland."[28]

3

BREAKING THE BOUNDARIES OF LIFE

T he wind was blasting sideways across the sharp ridge, each gust so full of moisture it felt not like air but nearly like an ocean wave. It knocked us cockeyed. With our boots slithering on slimy mud and eyes half-blinded by the mist, we could barely keep our balance. Ecologist Alan Pounds, a shiny ghost just ahead in his blue rain gear, was leading the way down Brillante Ridge in Monteverde Cloud Forest Preserve, Costa Rica. On this day, the place was certainly living up to its reputation as the focus of trade winds that career up and over Central America, depositing huge amounts of moisture in the form of mist and cloud. Pounds was on his yearly quest to find representatives of a species of distinctively orange amphibian in its restricted habitat along this gusty ridge. It seemed wet enough, the right place for the golden toad. But something has changed here, and the toads are now missing. Although about 1,500 were sighted in 1987, none have been seen since 1989. The toad's breeding pools remain empty, and it is feared extinct. This is an unpleasant first: a well-documented, peer-reviewed case of a species' disappearance because of climate change. The golden toad, *Bufo periglenes,* is probably gone forever; many more amphibians have not been seen recently; and other creatures in this rich ecosystem teeter on the edge.[1]

It is very clear to those who study individual species that climate changes are warping and tearing the living fabric of the planet. Living things must move or adapt—or they will die out—as average temperatures rise and precipitation and ocean currents shift. In polar regions this change is dramatic, and the same changes are at play in the mid-latitudes, from the ocean to the highest alpine mountains. Some changes may be barely noticeable to the casual observer, but studies of hundreds of plant and animal populations show an average movement toward the poles of almost 4 miles (6.1 km) per decade. As spring events occur ever earlier, the basic functions of life are shifting, such as bloom time and seed-set, mating and nest-building, and predator-prey relations. While most of these changes to other living things do not yet threaten humans, when what we eat is affected, starvation looms. In East Africa, warming is weakening the currents of Lake Tanganyika, which in turn slows the vertical mixing of water; with less cold, nutrient-rich water rising to combine with warmer surface waters, less plankton grows, and so young fish have less to eat. This is a threat to local people who get up to 40 percent of their protein from the fish catch. Creatures in many habitats are closer to extinction, and the golden toad appears to be just one of the species on the narrow edge of a wave that will change the distribution of life on Earth.[2]

An international group of nineteen ecologists, biologists, and climate scientists, after studying 1,103 species in a wide variety of habitats, predicted in 2004 that by midcentury up to one-third of these land plants and animal species may be pushed close to extinction. Climate change, they stated, is "likely to be the greatest threat in many if not most regions." Another study of European plants concluded that mountain plants will be especially sensitive to climate change; boreal species may fare better because they may be able to migrate northward in the face of warming. Research in Britain into the changing

PREVIOUS PAGE Porcupine caribou herd, Arctic National Wildlife Refuge, Alaska, crossing the Kongakut River on its annual migration. Migrating animals, which must synchronize their migration with the life cycles of the plants or animals they eat, are feeling the first effects of climate change. (JUNE 1997)

status of birds, butterflies, and plants over the last twenty to forty years also seems to point toward extinction. More than 70 percent of butterfly species are declining, and two species are no longer found in Britain at all. Of all 201 bird species known to nest in Britain, more than half have decreased in number, as have more than a quarter of native plants. Despite the gaps in the data and the scattered species that are increasing, these and other falling inventory numbers are "strengthening the hypothesis that the natural world is experiencing the sixth major extinction event in its history," the scientists wrote.[3]

These are dramatic words, and ones that challenge our easy assumption that life as we know it now is going on pretty much the same as it has for years, centuries, even millennia. After all, the last major extinction occurred a very long time ago when a meteorite slammed into the Earth and doomed the dinosaurs. Is global warming having that strong an impact?

There are signs that it may be, especially in concert with human-caused pollution and the destruction of rainforests and overexploitation of the oceans. The latest Global Biodiversity Outlook report, released in 2006 by the UN Convention on Biological Diversity, calls climate change "an increasingly significant driver of biodiversity loss." The studies highlighted in this chapter reveal strong impacts not only on individual species, but on entire habitats, including some of the most beloved and well-protected parks in the world. This research, representing hundreds of papers in the scientific literature, shows that other species on Earth are taking to their feet, or wings, or seed pods—however they move, they are moving. Nature is not uniform; not all species in a particular habitat are moving at the same time, or even in the same direction. But as the world changes as it warms, countless plants and animals have no choice but to follow the habitat they evolved into. However, some habitats will disappear, while novel climate zones appear. Many animals will be unable to migrate or adapt. We should take heed: the mountains, forests, meadows, and oceans being affected are also irreplaceable for their role in human life. It is our habitat, too.[4]

AMPHIBIANS DISAPPEAR IN MONTEVERDE

Cloud forests are sometimes called "fountain forests" because they are so important in the water cycle. They comb moisture out of fog, increasing by 15 to 100 percent the water available (as rainfall) to regions beyond

Atelopus varius, a cloud forest frog endemic to Costa Rica and surrounding Central American nations, appears to have shifted its range in response to regional climate changes. A study by herpetologist Alan Pounds and colleagues revealed that many indigenous amphibians have disappeared from rain- and cloud forest habitats in Central America, apparently as a result of new climate conditions that favor the growth of a deadly fungus. (FEBRUARY 1986)

the mountains. They are also fountains of biodiversity. Although montane cloud forests constitute only 2.5 percent of all tropical forests, their ready moisture contributes to a disproportionately wide variety of plant and animal life. Thirty percent of Peru's 272 endemic species, for example, live in cloud forests.

For many years, it was thought that tropical rainforests, lowland and montane, were essentially unaffected by climate change. Today, studies are showing that not only were they changed during past events like ice ages, but some areas are being affected right now by the warming atmosphere. The loss of the golden toad and other amphibians in Monteverde, Costa Rica, is particularly troubling because their habitats in the preserve are protected in the largely pristine montane rainforest. One cannot blame habitat loss—a major cause of amphibian declines worldwide. A recent global survey of frogs and toads showed that nearly all species listed as "possibly extinct" live in seemingly undisturbed habitats. Herpetologist Andrew Blaustein notes that half of the amphibians that have been the object of recent studies have been breeding earlier, a trend that correlates with evidence of global warming. Yet climate change, he points out, is but one factor in species disappearing; disease, pesticides, and wetland destruction must also be factored in. (Deforestation and ranching in surrounding areas, for example, may be affecting the Monteverde preserve.) Much study remains to be done on these sensitive creatures, wrote British herpetologist Trevor Beebee in *Conservation Biology;* nevertheless, enough evidence exists to say that global warming is a threat to some amphibians.[5]

When Alan Pounds began to contemplate reasons for the disappearance of the golden toad over just a few years, his first thought was that regional temperature increases were lifting the cloud level above the little amphibian's 6,500-foot (2000 m) home, depriving the cloud forest of moisture. Still, he knew the toad could not literally dry up in such a short time. Pounds suspected another culprit: an introduced chythrid fungus, known to attack the skin of amphibians, that had somehow reached epidemic proportions.

Herpetologists have been tracking mysterious amphibian die-offs for some years, including in Australia, where the same fungus was implicated. In Central America it was not just the golden toad that was disappearing. Scientific observations over the course of twenty years indicated the near-simultaneous disappearance of sixty-seven species of *Atelopus,* jewel-like forest harlequin frogs. These animals vanished not just at middling elevations, but also higher up and down through the lowland rainforests of Central America. Although the habitats varied, most of the losses occurred after particularly warm years. Warming and drying alone were not to blame, because these would have killed off the higher-elevation frogs before lower species.

Pounds and thirteen other biologists in the Americas and Japan set out to find out what other factor could explain so many amphibian extinctions. Their answer, published in January 2006, points to the unpredictable synergy that climate can have with disease. Careful cross-analysis of temperature, moisture, and optimum growing conditions for the chythrid fungus indicated that a warming climate favored the pest, allowing it to infect the skin of many frogs and kill them. This was especially true at elevations from 3,300 to 7,900 feet (1000–2400 m), where night temperatures rose and, during the day, warmer moist air increased cloudiness, protecting the fungus from hot sunlight. This was an ominous discovery. "We establish that global climate change is already causing the extinction of species," the researchers wrote. "We conclude that climate driven epidemics are an immediate threat to biodiversity."[6]

At Monteverde, many other changes seem to spring from climate change as well. Drier conditions in some parts of the cloud forest endanger tiny *Pleurothallic* canopy orchids. "We are now seeing two, three, even five days in a row without moisture," reports Karen Masters, a conservation biologist for the Council for International Educational Exchange (CIEE) at Monteverde. "This is very challenging to these orchids." Forest ecologist and canopy scientist Nalini Nadkarni warns that continued drying will drastically change the composition of the diverse epiphyte community that inhabits the cloud forest canopy. Meanwhile, recent repeat surveys of bats by Richard LaVal and of birds by Debra DeRosier (repeating a 1979 survey by Dr. George Powell, cofounder of the Cloud Forest Preserve) reveal that lowland, dry-habitat species are already moving higher into former cloud forest areas. Michael P. L. Fogden, in a survey of birds near the preserve headquarters, found that an average of nineteen lowland species per decade, including golden-crowned warblers, lesser greenlets, and keel-billed toucans, have been migrating into Monteverde as conditions change (and as some species react to deforestation in lowland Costa Rica). Shifts in insect, snake and lizard, and tree squirrel populations are also reported.[7]

EXTINCTION IN THE CLOUD FOREST

The cloud forests perhaps most at risk to lose a large number of species with only slight atmospheric warming are in the Australian tropical rainforests of Queensland. "These mountaintops are among the most endangered in Australia, and probably the world," said tropical biologist Steve Williams of James Cook University. "They're a World Heritage Area because they have these endemic species, which are remnants of Gondwana and are adapted to cooler conditions." The rainforest rises in several green waves directly from the warm waters of the Great Barrier Reef to about 5,200 feet (1600 m). "The forests themselves are very, very old," said Williams. "About four million years ago the entire northern part of Australia was rainforest. There was a forest here the size of the Amazon. As the continent moved north and became hotter and drier, the habitat contracted to this coast." Today this tiny group of mountain ranges is smaller than Connecticut—3,860 square miles (10,000 sq km)—yet it harbors Australia's most diverse flora and fauna. At the same time, it is a risky spot; seventeen plants, for example, are thought to have gone extinct since 1950, according to the World Wildlife Fund.

Williams is most concerned about the higher-elevation cloud forests, which are only a third of the total rainforest area but harbor eighty-six endemic vertebrates. These animals, which live only here— four ringtail possums, six microhylid frogs, and the golden bowerbird among them—"are restricted to above 600 meters [2,000 ft.] elevation, and there is nowhere for them to go," said Williams. Five of the frogs are found on single mountaintops, and the bowerbird is found only on several summits. The temperature at these elevations can be more than 27°F (15°C) cooler than along the coast. Meteorological records show that in the dry season, 40 percent of available moisture comes from fog capture by the tallest peaks as they are draped by air rising from the Coral Sea. If the cloud fails to form or forms

Queensland's wet tropics harbor Australia's most diverse flora and fauna in habitats such as the higher-elevation cloud forests, in which eighty-six endemic vertebrate species live. Tropical biologist Steve Williams and field assistants inventory a rare palm and fern forest at 3,900 feet (1200 m). Williams warns that four arboreal marsupials, six frog species, and a bower-bird are among the creatures that may not survive the rising temperatures that are drying these forests. [NOVEMBER 2005]

above the summits due to warmer air temperatures, the loss of this much moisture could be disastrous for some of the endemics.

Williams said he "never intended to work on climate change. I was always interested in biodiversity." His career includes hundreds of lengthy expeditions, during which he is as likely to dive out of his land cruiser to grab a python on the road as to set up transects for precise observations of marsupials and frogs. He knows the entire fauna by sight and sound; just the color of eye shine at night in a flashlight beam can enable him to identify a creature. But in 2003 he had a graduate student run a program to see what would happen to a set of data on mountain-forest biodiversity when certain variables, such as temperature, were tweaked even slightly. "The results were completely mind blowing," said Williams. "It changed the direction of my research immediately. Climate change was by far the greatest threat, far more than any other."

According to Williams's analysis, the eighty-six endemic vertebrate species will experience a 50 percent extinction rate even under relatively low estimates of inevitable twenty-first-century temperature rise. As the zone of fog moves up the mountain toward the summits, the habitable land area gets smaller, thus sharply reducing habitat size. Even if animals can move with the changing fog layers, there will be a lot less room for them. High temperature alone is sufficient to doom some arboreal marsupials, which cannot maintain their body temperatures if the air is hotter than 86°F (30°C). Reports are that ringtail possums overheat and fall out of the trees, dead. "It potentially takes just one day of six or seven hours of extreme heat, or two or three days of the hottest temperatures ever, and you could find dozens of dead possums on the ground."

Williams's realization of the great risk to so many unique species from as little as a two- to three-degree temperature change was a watershed moment for him, and for tropical biology in general. He described how he had been going to meetings about biodiversity and hearing all the stories of research hobbled by fragmented habitats and parks with porous boundaries, high levels of poaching, and poor enforcement. Here, in this World Heritage Area, "I felt quite smug having a [study site] that was completely protected and enforced." Unfortunately, with climate change, "it doesn't matter if it's a World Heritage Area."[8]

All over the tropics, big changes are being correlated with climate change. Hawaii has highly unique island flora and fauna, which are now under pressure from introduced predators and invasive species. Many of the most endangered birds have found refuge in the highlands of Hawaiʻi, Maui, and Kauaʻi, where more habitat remains and predation is lower. With rising temperatures, however, a very serious threat, avian malaria, is moving up in elevation. Studies on the slopes of Hawaii Volcanoes National Park show the malaria vector, the introduced southern house mosquito, to be a special danger to native honey creepers like the Iʻiwi, for whom the disease is almost always fatal. And as on other tropical mountains, as the mosquito zone rises up the slopes, the land area where the birds are safe shrinks.[9]

The observed warming this past century in Peru's tropical forest is ten times that of the warming it experienced coming out of the last ice age. A hallmark of tropical warming is that night temperatures increase faster than daytime readings. This may cause the tropical forests to give off more carbon dioxide than they absorb and help explain the more than 6 billion tons of carbon that was measured coming off the tropics during the 1998 El Niño, a very warm period. Plants not only perform photosynthesis (which takes in CO_2 and releases oxygen); they also respire like any living thing, and so give off CO_2. It was thought that these functions balanced or that the forests were a net absorber of carbon. But research at La Selva Biological Station in Costa Rica demonstrated that during warm nights respiration increases, while photosynthesis is shut down. This causes the trees to grow more slowly, further reducing carbon uptake. These results are not accepted by all tropical biologists, but it seems clear that climate changes are affecting the forests in ways not seen before. In parts of Amazonia that have remained undisturbed for more than 4,500 years, the rise in atmospheric CO_2 may be changing the composition of the forest. A twenty-year study of 13,700 trees in eighteen very isolated plots in Brazil, for example, has shown that some trees are now growing more slowly, whereas others are burgeoning un-expectedly, which could change the forest's total carbon storage.[10]

THE FIFTH HORSEMAN

The stresses on biodiversity—including human population pressure and development, overfishing, pesticides and other chemicals—are so plentiful that some scientists see an apocalypse on the horizon. But research shows that global warming is also a powerful horseman, and one that is spurring on separate changes. In the past few years, thousands of papers have been published on individual plants and animals showing strong correlations between the huge shifts that are beginning to threaten biodiversity and the observed direction of climate change. The research does not look at what is happening just today, in a single snapshot. All these species have backstories; they've been studied and observed for decades, with baseline knowledge going back thirty, fifty, even a hundred or more years. Many observations show that species' feeding habits, as well as their responses to heat, drought, and heavy rain, are changing under global warming. Current investigators have taken great pains to duplicate previous research designs, and to consider numerous possible reasons for the shifts they see occurring. They have also reassessed and combined these thousands of studies, as a way of viewing the big picture. They want to know whether there is an inarguable, coherent link between global warming and ecological change.

The broadest and most respected of all scientific synthesizers of research into the effects of climate change is Working Group II of the Intergovernmental Panel on Climate Change, which looks at impacts, adaptations, and vulnerabilities of natural systems. The IPCC, now in its fourth iteration, has just completed a review of research published since its third report in 2001. The IPCC review is the largest of what are called meta-analyses, considering many groups of data to detect shifts in nature and see if they can be attributed to the recorded rise in temperature. The two hundred scientists in Working Group II looked at over 650 studies of changes to the Earth and its inhabitants, which included 29,436 sets of long-term observations. They confirmed that effects of temperature increases are perva-

Boreal s
and tree
advance

Erosion and flooding
affect 184 of Alaska's
213 Native villages (2,

Warmer waters
affect walrus,
seabirds, salmon
(2, 3)

Forests attac
by insects;
fires increas

Oceans becoming
warmer, more acidic (4)

Lower plankton productic
affects ocean ecosystems

Western Antar
glaciers thinni
flowing faster

Extent of Arctic summer sea ice smallest ever measured, 2007 (1)

Greenland Ice Sheet melting with increasing speed (1)

ar bears starving, eatened by low sea ice (2)

More methane released by thawing tundra (2)

Eurasian growing season 18 days longer than in 1981 (3)

Less snow cover throughout Northern Hemisphere (4)

Deadly heat wave, 2003; Europe experiences highest temperatures in 500 years (4)

Warming causes both more intense storms and more extensive droughts (4)

ricultural zones e northward (3)

Climate linked to changing salinity and currents (1)

astal cities reatened by sing seas (4)

Birds shift migration routes and advance breeding times (3)

Asian cities choke in climate-induced dust, smoke (3)

Warmer sea surface causing more intense hurricanes (4)

Africa suffers highest temperatures on record, 2004–2005 (4)

Animal habitats everywhere shifting poleward; ecosystems are disrupted (3)

hibians pushed xtinction by climate ts (3)

Asthma, mosquito-borne illnesses increase (4)

Tropical forest tree growth altered by climate changes (3)

Sustained aridity in southern Africa (4)

Sea level rising twice as fast as in twentieth century (4)

Coral reefs experience severe bleaching events (4)

Mountain glaciers shrink worldwide (1)

Atmospheric CO_2 levels highest in 800,000 years (5)

tarctic Peninsula mperature 5°F higher an 50 years ago (1)

Penguin range changes as ice and krill decline (2)

Extreme weather events and changes in animal behavior are transforming the world, as increasing greenhouse gases and rising atmospheric temperatures take their toll. Numbers in parentheses indicate the chapter where each event is discussed.

sive across the globe and that, in the words of Cynthia Rosenzweig, a NASA scientist who took part, "the changes have intensified." They tallied ongoing habitat changes, landscape alterations, and loss of biodiversity on every continent. A feature of the 2007 report is a scaled estimate of the changes expected as temperatures rise, plants and animals experience increasing extinctions, and hundreds of millions of humans suffer from coastal flooding, drinking water shortages, and disease. The severity of the effects are dependent on how much we can control greenhouse emissions, and thus global warming. The 2007 IPCC report echoes independent metastudies completed during the past five years. Another analysis of 866 peer-reviewed studies (40 percent of them published in the past three years) showed "a clear, globally coherent conclusion: Twentieth-century [human-made] global warming has already affected Earth's biota."[11]

At my own home near Portland, Oregon, freezing days are fewer, more plants (and weeds) over-winter, and the first Pacific tree frogs sing earlier—in 2005 and 2006, almost a week earlier than in my previous thirty years of casual observation. Gardeners don't need scientists to tell them spring is coming earlier and frost later, of course. But neither is it an urban myth or lucky almanac prediction. The National Arbor Day Foundation has had to adjust the hardiness zones on its new U.S. tree-planting map, moving them many miles north relative to the 1990 version. An animated feature on the foundation's website shows northern Nebraska changing from zone 4 to 5 and states from Mississippi through Georgia becoming almost entirely zone 8. Satellite and surface weather station records prove that the growing season is eighteen days longer in Eurasia and twelve days longer in North America compared to 1981. Trees all over Europe are leafing out almost eleven days earlier than in the early 1960s, as measured in an international network of botanical gardens. Digging deeper into 125,000 records of the timing of yearly springtime events—a study called phenology—European scholars discovered that 78 percent of such events had moved earlier in the year since 1971. A significant number of these changes match "measured national warming across 19 European countries." Across the entire Northern Hemisphere, records that date back into the fifties confirm that the onset of spring warmth, as well as first leafing and blooming dates, is uniformly sooner. In North America, the pioneering American ecologist Aldo Leopold kept meticulous records of natural events at his Wisconsin "shack" through the 1930s and 1940s, including the arrival of birds, onset of wildflower bloom, tree leafing events, and formation and breakup of lake ice. Now his daughter, botanist Nina Bradley, has matched fifty-five of the spring events of 1936–47 to ones she and colleagues recorded during the 1980s and 1990s. About half occurred earlier in the season, the overall trend being that over the course of sixty years, spring advanced 1.2 days per decade in Wisconsin. In the Washington DC area, meanwhile, a survey of 100 flowering plant species—including the Japanese cherry trees along the Tidal Basin—showed earlier blooming times (in one case, by forty-six days) in eighty-nine species over a twenty-nine-year period.[12]

Birders and amateur naturalists are also documenting climate-based changes. In Britain, their records are accepted as part of the history of British nature, and in the United States, too, such events as the Audubon Christmas Bird Count contribute valuable information. Elizabeth Losey of Germfask, in Michigan's Upper Peninsula, was one such amateur birdwatcher. In 1947 she became the first female field biologist in the USFWS, stationed at the Seney National Wildlife Refuge. For fifty years, in the midst of performing the administrative and habitat protection duties of her position, she kept personal bird migration records. She held on to them, except for a few of her earliest notebooks, which, she said, were tossed out by her overly tidy husband. When she looked back at the carefully written dates and species, she "was interested in the numbers. The decrease in numbers. Some species we don't even have anymore." Talking with me in 2000, at age eighty-seven, Losey seemed nostalgic about her days on the refuge. She had seen gross changes in the landscape. "It's a far different place than when I was there forty years ago. Waterfowl habitat is way down; we just don't have it anymore. Cattails have moved in, brush has moved in. There was more open water." In 1999, ornithologist Terry Root, then of the University

Purple deadnettle *(Lamium purpureum)*, blooming an average thirty-nine days earlier over a twenty-nine-year period, is among the many flowering plants observed in a study conducted near Washington DC. Eighty-nine of one hundred familiar plants—including the Japanese cherry trees along the Tidal Basin—showed earlier blooming times. (JUNE 2001)

An early-migrating chiff-chaff warbler, caught by David Walker in mist nets at England's Dungeness Bird Observatory, is representative of the thousands of species of birds migrating and nesting earlier or adjusting their ranges under pressure from the effects of global warming. Long-term records kept by both amateur and professional naturalists are proving that more than half of all living things studied are already reacting to a warming climate. (APRIL 2001)

of Michigan, compared Losey's records to current figures and found that forty-six bird species arrive earlier in upper Michigan now—some by as much as twenty-one days—while only one arrives later. In Britain, when Humphrey Crick and colleagues at the British Trust for Ornithology analyzed almost 75,000 nesting records for sixty-five species of British birds over twenty-five years, fifty-one species were nesting earlier (by 8.8 days on average), while fourteen were found to be nesting a bit later. Higher spring temperatures correlate directly with earlier egg laying by flycatchers across Europe.[13]

The trend is clear to the Wildlife Society, whose 2004 review of North American wildlife changes concluded: "The overall ranges of many bird species are now thought to be as much influenced directly by climate as by availability of particular habitats." Jeff Price of the American Bird Conservancy says the effect is greatest on the tiniest migrants, like Cape May, bay-breasted, and blue-winged warblers, which "appear to be more sensitive than other North American birds to temperature and climate changes." He also cited the ovenbird and redstart, which are no longer found in the southern Appalachians. Rufous hummingbirds are sighted across the southern United States now in winter, rather than being confined mostly to Mexico, according to records compiled since 1900.[14]

Not all birds are moving north or arriving earlier. Birds studied in the French Alps, for example, have not changed their altitudinal range despite regional warming of 4°F (2.2°C). Yet regardless whether the birds are moving in response to temperature changes, they may well be finding less of the insect food they need for migration and breeding, because the bugs are changing in their own way. Pied flycatchers arriving in the Netherlands from Africa, for example, rely on caterpillars to feed their young. Since 1985, some flycatcher populations have declined by up to 90 percent because the peak period of abundance of their larval prey now occurs up to sixteen days earlier; there isn't enough food left when the chicks really need it. In England, ornithologist Andrew Gosler of Oxford University monitors the nesting of hundreds of pairs of great tits (a relative of chickadees) in a study that has documented a three-week advance in egg laying since the 1960s. Gosler considers this "entirely consistent with the weather," though he notes that now the birds seem "more inclined to be out of synch with the caterpillars." In a different population of great tits in the Netherlands, some birds have apparently begun adapting to temperature changes, and may be passing on the trait of earlier nesting. The Dutch study is one of only a few that show any long-term adaptation to rapid climate changes.[15]

These findings all hint that ecosystems are not shifting in one piece, as we might expect from looking at a map of the Earth with neatly colored habitat zones. Rather, it appears that habitats as we know them are going to fragment, as each part, like the birds and insects and seasonal weather, responds individually to climate change.

PLANTS MOVING UP THE MOUNTAINS

In summer 2004, a crew of six researchers and an ever-shifting cast of family members assembled in a tent camp a day's walk above the village of Gries in Tirol, Austria. Daily, the botanists trooped up Schrankogel Mountain, past dairy cattle and through a landscape worthy of *The Sound of Music,* to reach their alpine study sites. The highest are steeply inclined patches of earth on a rocky ridge more than 11,000 feet (3400 m) high. The scientists were here to reinventory 380 of more than a thousand

Botanist Harald Pauli, who with colleagues from the University of Vienna set up a plant study on Schrankogel Mountain, Austria, in 1994, counts individual plants and charts their locations ten years later. The study has found that most plants are moving up the mountain, and that as temperatures rise the alpine species are losing ground. (AUGUST 2004)

Biologist Camille Parmesan found in 1996 that the North America Edith's checkerspot butterfly had moved north by about sixty miles and up the mountains by about 300 feet in the twentieth century. Collaborating next with scientists in Europe, she found that two-thirds of butterfly species on that continent had shifted northward by 22 to 150 miles. (JUNE 2002)

DETECTING IMPACTS OF CLIMATE CHANGE

CAMILLE PARMESAN

University of Texas

Things change all the time. Suddenly I see masses of white winged doves in my backyard in Austin. Is this a sign of global warming? Maybe, but humans are doing a lot more than changing the climate. We dominate the land, and we are beginning to dominate the sea. We know that habitat destruction, creation of urban landscapes, pollution, pesticides, and roads all affect where plants and animals can or cannot live. Then there's the element of chance. A population of pikas might go extinct because a landslide ripped across the rocky slope where they lived. So life is in flux, even without humans. How can scientists tease out the effects of global warming from other human or natural forces that influence wildlife? It can be done, but it requires the skills of a detective as much as those of a scientist.

The first step is to simplify the question. Rather than look for all conceivable impacts of climate change, we look for predictable and observable responses. If wildlife is indeed responding to recent warming, we can predict that species' ranges will move toward the North and South Poles and up mountainsides, just as they did when the glaciers retreated some eighteen thousand years ago. Global warming should also alter the timing of spring. Migrations, nesting, and flowering should be happening a bit earlier each year. We have seen these signs. Thousands of amateur naturalists have recorded the rites of spring in painstaking detail and witnessed their gradual shift earlier over the decades. Likewise, records of where plants and animals live, going as far back as two hundred fifty years, reveal that during the past twenty years butterflies and birds have moved northward and trees have moved up mountainsides, relative to where they were one hundred or even fifty years ago.

If a handful of species does this, it could be coincidence. If dozens of species do this, it suggests a broader pattern of change. In fact, we've found that, in concert with gradual changes in temperature and rainfall, hundreds of species have either shifted where they live or when they start spring activities. In a study that synthesized data from more than 1,500 species worldwide, we estimated that more than half have responded to recent warming trends. Such numbers make it nearly impossible for the connections between changes in the natural world and greenhouse gas–driven global warming to be dismissed as "coincidence."

But that's not the end of the story. Biologists have spent decades studying quirky details of how plants and animals live. Until the rise of a new moral ethic in the late 1970s, some biologists performed horrid experiments in which they stressed

Parmesan traveled to every location mentioned in these records to ask a simple question: "Is it here now, or not?" She found that the insect had moved north; she found, too, that it was four times more likely to die out at the far southern end of its range (Mexico) than at the northern end (British Columbia). In the southern part of the range, the annual plants on which checkerspots lay their eggs are very sensitive to drought. Because of warmer, drier conditions in recent years, when the eggs hatch, the plant has often already matured and started to set seed, providing sparse food for the caterpillars. A whole generation of caterpillars thus starves, and the cycle of reproduction ceases. At the other end of the insect's range, in southern British Columbia, in contrast, the warmer climate means that more butterflies survive to lay eggs and create new populations. The range had shifted northward by about 60 miles, and it had moved up in elevation by about 300 feet. Parmesan calculated that this amount of shift exactly matched what you'd expect from the 1°F (0.6°C) of warming that North America experienced in the twentieth century. Her research, published in *Nature* in 1996, was the first to provide evidence of a creature changing its range due to climate.[31]

Next, Parmesan entered into a collaboration with a group of European lepidopterists, who gathered extensive records on fifty-seven butterflies dating back to the 1700s. They found that two-thirds of these

an animal—making it colder or hotter—until it died. These experiments often documented exactly how biological processes were affected as stress increased. Long-term physiological research on plants and animals has allowed modern biologists to realize that we are now seeing behavioral responses to gradual warming of the globe and even to know precisely *why* and *how* certain changes in weather patterns make conditions for wild species more or less stressful, as the examples below show.

Combining what we know about the basic biology of the species we study with long-term observation completes our detective work. Consider the pika. This tiny montane mammal needs to eat many times a day, but it won't forage for food in temperatures above 20°C (68°F), and caging experiments that forced animals to be out in the open on hot days showed that they die if placed in air temperatures exceeding 31°C (87.8°F). Pika populations are going extinct at lower elevations, be-low about 8,000 feet, where temperatures, on average, have warmed. Pikas already live as high as the mountains go, and thus they and other "cold earth" species already dwell at the extreme limits of our present "warm earth" climate. They literally have nowhere to go as Earth continues to warm.

My own study species is a delicate butterfly, Edith's checkerspot. The population health of these butterflies is strongly driven by temperature, rain, and snow. In one montane population, a very warm winter caused butterfly emergence to be out of synch with the flowering of its nectar plants; there was no food when adults emerged. The hillside was littered with bright orange jewels—not the blossoms of flowers but the bodies of butterflies, starved while their wings were still soft after metamorphosis. In general, false springs (which occur when a snowpack is light and melts early but winter storms may still come on) have caused the extinctions of many mid-elevation mountain populations. In a very low elevation population, hot temperatures in late spring caused the tiny annual host plant to dry up before the caterpillars were big enough to enter their summer sleep, so the caterpillars starved. In the past one hundred years, as winters have become warmer, summers have become hotter, and spring has come earlier, Edith's checkerspot has moved its range gradually northward, shifting out of Mexico and blossoming in Canada. This butterfly's range has also moved upward into the Sierra Nevada.

Coincidence? Not likely, when we see the same story repeated over and over, from Alaska to the Antarctic, from the Great Barrier Reef to the North Sea, from the cloud forests of Costa Rica to the rolling hills of England. There is now a scientific consensus, born from more than one hundred years of observing and studying the biology of wild species, that global warming has already had an impact on the natural world. ✦

species had shifted northward by 22 to 150 miles, again in a pattern consistent with temperature increases in Europe. No butterflies were found to have shifted south. Parmesan also identified many locations in southern France with apparently intact food-plant habitat but no butterflies, such as the beautiful Apollo (*Parnassius apollo*). Working in and around Mt. Aigonal in Les Cévennes National Park, a rugged landscape of pine and beach forests, she has recently launched a new study of butterflies that will try to explain why species in the same habitat have very different responses to regional warming.[32]

Now an associate professor in the Department of Integrative Biology at the University of Texas, Parmesan has cowritten a series of papers analyzing thousands of other changing animals and plants, to show that the consequences of climate change are already clearly upon us. "Species are reacting to climate change," she told me. "That's established. It's not controversial." In conservation biology, she said, we need to know what to conserve and learn "to make climate-predictive models that more accurately reflect biological reality." In other words, we need to try to understand what the animal experiences, and what is really crucial for each species' survival. "What we find," she said, "is that predominately, the changes we're seeing are changes that were predicted from the climate change reports twenty to thirty years ago. So it does look as if the climate change of the twentieth century, which is relatively small, has impacted an awful lot of natural systems . . . everything from birds to butterflies, all the way through trees and even into marine systems."[33]

Despite all this scientific evidence, some policy makers and doubters have continued to challenge its validity. In some cases, individual data sets have been questioned. Were they, for example, large enough? Were other obvious interpretations overlooked? A more general criticism is that observational studies of changes in wildlife and habitat are not scientifically rigorous because they are not controlled experiments (that is, with two populations of subjects, one that is directly influenced by the variable under investigation and one that is not). Although these studies do indicate a relationship with temperature, skeptics argue that that variable may not be the operative one. Some critics point out, too, that observational studies cannot directly tie population or habitat changes to human emissions of greenhouse gases.

Controlled experiments with plants are actually quite common, including warming plants and wafting extra carbon dioxide gas over them to see what they do. These studies usually find plants changing as predicted by plant physiology and the general theory of global warming. Animal studies, however, typically involve a large number of animals observed over a long period of time to see if they react (or not) to climatic changes in their particular habitat. Although scientists take great pains to maintain consistent observation techniques and remain alert for bias, there is no way to have absolute control with wild creatures. Still, methods do exist for judging the reliability of inferred correlations. Parmesan explained, for instance, that in her most recent papers she used a "broader method of hypothesis-testing termed 'scientific inference,' which is common in astronomy and physics where direct experimentation is often impossible. The basic idea is that you use information from multiple, independent sources and come up with conclusions based on the consistency of answers when the problem is looked at from many different angles." The influence of global warming tied to human CO_2, for example, should leave a "fingerprint" of effect across a wide array of species and habitats. Parmesan, working separately with economist Gary Yohe and ornithologist Hector Galbraith, analyzed data on more than two thousand species, including ones that weren't showing any signs of change. They also investigated more than one hundred years' worth of detailed experimental work on temperature tolerances of different species, to better understand what these species could be expected to do in the face of climate change. The scientists went so far as to recompute the data from some studies, just to make sure initial conclusions were correct. In North America alone, forty long-term studies have pointedly associated climate change with ecological impacts on ecosystems ranging from tide pools to alpine meadows. Despite the diversity of subjects, the scientists concluded that more than half these studies contain "strong evidence of a direct link" to recent rapid temperature changes. The most striking finding was that wild species worldwide show "significant range shifts averaging about four miles (6.1 km) per decade towards the poles (or movement upslope in the mountains). Spring events are earlier by 2.3 days per decade."[34]

However, some critics of global warming research charge that these studies do not quite get us to the point of showing that human activities are causing all these individual changes in the natural world. Could wildlife behavior, which we know is strongly tied to climate and temperature within a habitat,

CHALLENGES TO BIODIVERSITY IN A CHANGING CLIMATE

THOMAS E. LOVEJOY

H. John Heinz III Center for Science, Economics, and the Environment

In the 1980s the director-general of the United Nations Environment Programme (UNEP), Mostafa Tolba, invited a small group of scientists and environmentalists to Nairobi, Kenya, to advise him on what ultimately became the Convention on Biological Diversity. Having studied the link between climate change and the diversity of life, I warned: "You had better pay attention to climate change, or you can forget about biological diversity." While that warning is somewhat of an overstatement, the essence is true, because climate change and the natural world are linked inextricably in complex ways.

That understanding was based on research that Rob Peters and I subsequently published as *Global Warming and Biological Diversity* in 1992, when the Earth Summit initiated conventions on both biological diversity and climate change. Thirteen years later, the successor volume, *Climate Change and Biodiversity*, edited with Lee Hannah, demonstrated a distinctly changing picture.

The primary difference is that nature can now be seen responding to the human-caused climate change that has already taken place. There are changes in timing—flowering, nesting, and migration dates. Some species have been clearly documented as moving—upslope or poleward on the land, and in equivalent ways in the seas—to track their required climatic conditions.

Climate change is hardly new in the history of life on Earth. Indeed, it is a given. We know about the ebb and flow of glaciers in the Northern Hemisphere during the last 100,000 years. So to some extent we can use knowledge of past climate change to help anticipate what future climate change will mean for nature. One thing we know with certainty from paleoecology is that biological communities do not move as a unit. Rather, individual species move independently in both direction and rate, causing ecosystems to disassemble and species to regroup in novel ecosystems.

This presents an enormous challenge for the design and management of protected areas. Today these are based squarely on the location of current ecosystems—on a static vision of nature. Besides having biologically functional landscapes to protect diversity, it is now also essential to maintain physical connections between different landscapes. This connectivity will provide flexibility for organisms responding to climate change, especially given the obstacle course of human-modified landscapes. More than ever it is important to think of ourselves as living *within* nature rather than confining nature to isolated, protected patches in a landscape dominated by human activities.

Climate change poses a particular threat for ecosystems that lack options for reducing its effects. Species at the top of mountains, for example, have nowhere upslope to move as temperatures rise, so many will vanish into thin air. Similarly, species on low islands vulnerable to sea level rise may be marooned, having nowhere to disperse. Such ecosystems cannot "adapt naturally" (the language of the climate change convention). The only way to conserve species in these ecosystems is to limit the extent of human-induced climate change.

Reducing the total stress on ecosystems will make them more resilient to climate change. For example, tropical coral reefs, subject to coral bleaching with just small increases in water temperature, are also vulnerable to other environmental stresses, especially sedimentation due to deforestation and other agricultural practices on adjacent land. Eliminating those stresses may enable coral reefs to better withstand bleaching events.

Another worry for corals and other ocean life is higher acidity of seawater as a direct consequence of the increase of carbon dioxide in the atmosphere, because some of the carbon dioxide absorbed becomes carbonic acid. This threatens to upset the calcium carbonate equilibrium so essential to tens of thousands of species of marine organisms in their building of skeletons or shells. It could also disrupt the carbon-storing capacity of the oceans, leading to even higher atmospheric concentrations of carbon dioxide.

Of the three main concerns of the climate change convention—agriculture, economics, and ecosystems—clearly the natural world is the most sensitive. Present and projected effects on biological diversity argue powerfully for a much more immediate and concerted effort to stabilize atmospheric composition and to limit the extent of human-generated climate change. ✦

be directly linked to the unquestioned fact that increasing greenhouse gases raise atmospheric temperatures? To address this issue, a 2005 study asked if plants and animals themselves could be used as "thermometers." Four scientists at Stanford started with documented observations of how 145 species in North America, Europe, and Asia actually changed in terms of earlier blooming or mating time, northward or upward shifts in range, and other behavior. For each species they then compared these results with what would be expected—over the same time period as the observations—from only natural atmospheric changes, using a computer simulation with no added greenhouse gases. They then recalculated against known human gas emissions only, and again against natural and human greenhouse gases combined. In 92 percent, or 133, of the species surveyed, the best fit was the combination of natural and human forces on the atmosphere. This sample of the natural world shows that humans, indeed, are causing life to react to climate shifts and habitat changes. In 2008, a similar "joint attribution" analysis of the IPCC's 29,400 data sets (page 84) correlated 90 percent of those changes with human emissions. Coincidentally, this work confirms the accuracy of the frequently quoted average world atmospheric readings, which also could only have risen as much as they have since 1970 with the addition of the measured human greenhouse emissions.[35]

TINY ARKS BUFFETED BY CLIMATE CHANGE

As the climate shifts, plant and animal species are beginning to move, each in its own way and at its own pace, in order to stay within unique, evolved envelopes of temperature, moisture, and light. A cen-

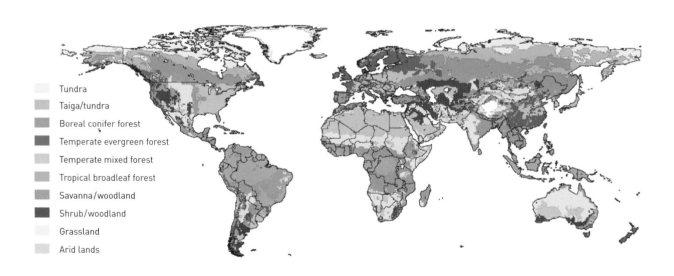

Tundra
Taiga/tundra
Boreal conifer forest
Temperate evergreen forest
Temperate mixed forest
Tropical broadleaf forest
Savanna/woodland
Shrub/woodland
Grassland
Arid lands

The Earth's recognized vegetation zones, or habitat areas, are shifting and fragmenting. As average temperatures rise, some forested areas are becoming shrubland, for example. These changes affect the animals—including humans—that have adapted to these habitats. [ADAPTED FROM NEILSON AND DRAPEK, *GLOBAL CHANGE BIOLOGY* 4 (1998): 505–21]

tral tenet of biology and ecology is that living things are inextricably tied to their habitat. As changes proceed, many species will be able to shift their range or alter their behavior in response. But some will be blocked because they've run out of mountain or island, because potentially suitable habitat has been destroyed, or because they can't move fast enough. What will become of the familiar habitat associations of Earth or the biodiversity and genetic flexibility they contain?

Unfortunately, ecosystems are unlikely to stay intact. "Communities of species do not move together," according to conservation biologist Thomas Lovejoy. "[Species] move individually at different rates and in different directions." As climate change intensifies, the life zones and ecological associations familiar to us from introductory biology courses, represented by multicolored bands splashed across world maps, are not going to move in synchrony. Rather, they will deform unevenly as the plants and animals within them react in varying ways. No less than the ice shelves of Antarctica and the permafrost of the Arctic, ecosystems worldwide are rending and disintegrating. With this, the rich biodiversity of Earth, the flow of life that humans rely on, is threatened.[36]

A major factor in this disintegration is the range of speeds at which species adapt to ecosystem changes. Looking back, by reading ice cores, pollen layers, and animal remains, scientists have been able to measure the rate and process of migrations as the glaciers retreated following the last ice age. During those thousands of years, according to a 2005 volume titled *Climate Change and Biodiversity*, edited by Lovejoy and Lee Hannah, a senior fellow at the Center for Applied Biodiversity Science, a few very rapid atmospheric temperature changes occurred. Those rapid and large climatic changes, notes Brian Huntley in *Climate Change and Biodiversity*, "triggered substantial numbers of extinctions." Hardier, or better placed, creatures survived only by embarking on extensive migrations as some habitats shrank or disappeared, while others grew. Adaptation was not an option under such volatile conditions. Adaptation, rather, requires long periods of very slow change, allowing animals to adjust their physical behaviors and characteristics in a process that leads eventually to the evolution of new species. As is pointed out in *Climate Change and Biodiversity*, such slow changes—on the order of 1°F (0.6°C) over a thousand years (compared to the one degree rise over the last hundred years)—were common in the distant past. In recent millennia, leading up to the present day, the climate has been unusually warm and stable. That stability, however, seems to be gone. This means animals and plants will need to be exceptionally flexible or remarkably mobile if they are to survive. If by flexibility we mean the ability of species to continue to live in a habitat that is changing, Camille Parmesan's research and analysis indicates that although there are always extraordinary individuals that can withstand stresses, "there is no evidence for [a] change in the absolute climate-tolerance of a species." Unfortunately, species usually go extinct.

Free-living bacteria, plankton, and small insects and plants—parts of the crucial bottom of the food chain—may be able to move (perhaps even evolve) relatively quickly. This is seen in the northward movement of plankton around the British Isles, for example, and in an apparent genetic adaptation to longer summers in a species of eastern North American mosquito. But larger organisms, especially plants and trees that have evolved into narrow ecological niches over thousands of years, may not be able to change fast enough and could well die out. In a report for the World Wildlife Fund, Jay Mal-

The Great Barrier Reef, Australia, off Cairns, Queensland, was one of the first sites to be designated a World Heritage Area. Along its 1,200 miles (2000 km) are 2,800 individual reefs harboring more than 5,500 species of fish and mollusks. It is threatened as never before by dangers—rising sea temperature and acidity—brought on by greenhouse gas emissions. (FEBRUARY 2005)

colm of the University of Toronto's department of forestry and ecologist Adam Markham estimated that today, global warming is pushing plant species to migrate at a rate ten times greater than rates recorded from the last glacial retreat. Migration of eastern forest trees appears to have occurred at a rate of more than half a mile (1 km) a year, according to forest ecologist Louis Pitelka. However, he wrote, this does "not guarantee rapid plant migration in response to future climate change." For one thing, people have greatly altered landscapes," and many of the resulting blockades and gaps impede migration. Moreover, Pitelka stated, the species whose past migrations we know of are not necessarily representative of the surrounding plant community as a whole. He warned of "the unpleasant prospect that, in the event of rapid climate change, unwanted species—weeds—would be the species that would have little trouble shifting their ranges."[37]

Investigations of sensitive ecosystems like the tundra and California grasslands show that changing the balance of temperature, CO_2, and precipitation reduces diversity. Plants central to the diet of grazing animals and insects may not thrive in the new conditions. In extreme cases, an ecosystem could shift permanently, taking on totally different characteristics, under stress of rapid climate change. It is a well-known ecological process, such as when a lake becomes a bog or when a forest fails to regenerate after a fire or logging. Aquatic ecologist Marten Scheffer and colleagues warned in 2001 that ecosystems from coral reefs to deserts already have problems from disease and damage from storms or by humans. A severe climate event could rapidly overcome a system's weakened resilience. Parmesan says in her 2006 paper analyzing 866 studies that "more crucial than any absolute change in timing of a single species is the potential disruption of coordination in timing between the life cycles of predators and their prey, herbivorous insects and their host plants, parasitoids and their host insects, and insect pollinators with flowering plants." "This," say Lovejoy and Hannah, "implies that vegetation communities may be torn apart and reassembled in novel ways."[38]

The implications are enormous. While ecosystems are undergoing great stress, they are more and more important as suppliers of crucial biological services and as repositories of the world's genetic pool. Think especially of the rich lands of Latin America, like the Amazon, and of Africa, which harbors about one-fifth of all plants, mammals, and birds. In a world where 80 percent of the land is influenced by roads, habitation, river traffic, or agriculture, according to the Human Footprint Project, where overfishing and pollution have reduced many fish populations, where 6.5 billion humans (rising to above 9 billion in this century, if estimates hold) vie for space and returns from the Earth—in such a world the need for the rich benefits of natural landscapes becomes ever more pressing. Yet only about 12 percent of the Earth's land, and much less of the ocean, is under some kind of protection, and many of the areas that have been set aside are, whether for political or social reasons, too small to support the wildlife and natural features they were intended to protect. Some were founded not on ecological values, but rather for scenic or historical or economic reasons. As Jeff McNeeley, chief scientist of the World Conservation Union (IUCN), stated at the 2005 Stony Brook World Environmental Forum on parks and reserves, they are not at all separate from human life. "Protected areas mean culture. We have created them." Although we think of such lands as storehouses of biodiversity, many are crucial to local people for water, building materials, hunting, recreation, and spirituality. Protected areas, in other words,

RISKING THE GIFTS OF THE EARTH

Reserves such as the Great Barrier Reef, Monteverde, and the Arctic Refuge are important not just for what they contain but also for their connections and contributions to the surrounding regions and the life of humans. The same can be said for a well-tended pasture and verdant oasis. All living things, from microbes to humankind, are dependent on the ecosystem services produced by the atmosphere, land, and oceans of the Earth. In the words of the UN Convention on Biological Diversity, the planet's natural functions provide "the goods and services that are crucial for human survival and well being."

The convention organizes these services into those that support basic life needs; those that regulate the climate and water of earth; those that provision us with food, shelter, and other resources; and those that are the basis of cultural assets. These services are threatened by rapid climate change. Some examples follow.

SUPPORTING SERVICES
Soil formation
Nutrient cycling
Oxygen/carbon cycling
Primary production by plankton and algae

Rising temperatures may dry out or burn soils and inhibit their creation. Warming and other weather changes can affect the ability of plants and animals to absorb nutrients. Plants do not uniformly absorb extra CO_2 under the stress of heat and lack of basic requirements, such as water. In the ocean, phytoplankton and algae at the base of the food chain take up huge quantities of CO_2 and produce new organic matter in a process known as primary production, but heat and acidification from additional CO_2 appear to inhibit this production.

REGULATING SERVICES
Climate regulation
Water cycle, including water purification
Flood and storm protection
Waste disposal
Pollination
Nitrogen fixation
Control of pests and diseases

These processes are disturbed by changes in carbon dioxide and methane levels in the atmosphere, rising temperatures, loss of glaciers, thawing permafrost, changes in weather systems, shifting of seasons and habitats, rising sea levels, and higher ocean temperatures and acidity.

PROVISIONING SERVICES
Accessible fresh water
Plants and animals for food, clothing, tools, fuel, building materials, pharmaceuticals, biotechnology, and other uses

This is where climate shifts and greenhouse gases can do the most damage to ecosystems. Fresh water depends on glaciers, forests, and natural soil filtering, all of which suffer under global warming. Higher CO_2 levels are causing some plants, including weeds and pathogens, to grow faster, while in other places crops and wild food yields are diminishing as a direct result of climate disruptions. Ocean fisheries and seabird populations crash when nutrient-rich currents and upwellings change. More frequent forest fires, especially in the tropics, can lead to loss of food and medicinal plants. Damaged protected areas or natural refuges can no long harbor crucial biodiversity. Decreased biodiversity portends fewer sources of new drugs and biotechnology.

CULTURAL SERVICES
Cultural diversity
Knowledge systems
Inspiration, spirituality, religious values
Nature appreciation
Recreation

The social value of the Earth's services is impossible to calculate, as are the impending losses through rapid climate change. Over millennia, humans have developed myriad cultures in response to the terrain, weather, flora, and fauna that exist in the places they have settled. Thus, the diversity found in nature is in large part responsible for our cultural diversity and for the knowledge systems that each culture has refined over time. Nature shapes our aesthetic sense, gives form to the symbols we find most potent, and is the font of our spirituality. This may be especially apparent in indigenous communities, but urban dwellers in industrialized nations also depend on natural settings as places for relaxation, recreation, and renewal. Beyond these considerations, many people believe that all the services are God's creation, a gift that it is profoundly immoral to misuse or damage.

These services have enormous economic and social value even if they are not traded and carry no price tags. (One attempt at gauging the value of replacing seventeen natural services, from waste recycling to recreation, arrived at a figure of up to $54 trillion per year.) Human technology alone is insufficient to duplicate these services. Although some are part of the economic system, most are used daily without thought or acknowledgment; others are seen as part of the world commons, for all to use freely.

A worldwide inventory, the Millennium Ecosystem Assessment, has found that approximately 60 percent of ecosystem services "are being degraded or used unsustainably." Most of the damage has been done since the mid-twentieth century as a result of overuse of the land, overfishing, and industrial pollution. Climate change exacerbates these abuses and brings new dangers.

Sources: Classification by Convention on Biological Diversity; list adapted by Gary Braasch with additions from Millennium Ecosystem Assessment and other sources; Secretariat of the Convention on Biological Diversity, "Interlinkages between Biological Diversity and Climate Change: Advice on the Integration of Biodiversity Considerations into the Implementation of the United Nations Framework Convention on Climate Change and Its Kyoto Protocol," CBD Technical Series no. 10, Montreal, Oct. 2003, pp. 1–2; Millennium Ecosystem Assessment, Ecosystems and Human Well-being: Synthesis (Washington, DC: Island Press, 2005), pp. v, 2, 6; Robert Costanza et al., "The Value of the World's Ecosystem Services and Natural Capital," Nature 387 (15 May 1997). See also Yvonne Baskin, The Work of Nature: How the Diversity of Life Sustains Us (Washington, DC: Island Press, 1997); and Gretchen C. Daily, ed., Nature's Services: Societal Dependence on Natural Ecosystems (Washington, DC: Island Press, 1997). Chalon Emmons provided valuable insight into cultural values.

are part of the habitat for human life on Earth, because they help guard the natural world we all rely on. And in a time of global warming, scientists have told the UN Convention on Biological Diversity, "genetically diverse populations and species-rich ecosystems have a greater potential to adapt to climate change."[39]

Perversely, in many cases rapid climate change is knocking the ecological foundations of the parks from under them. If climatic and life zones move outside a park's constrained boundaries, little hope remains for the associated animals, plants, and watersheds. This includes the hundreds of migratory birds and other animals that rely on specific habitats during their yearly journeys. "We can think of our entire nature reserve and park system as a static network," wrote Louis Pitelka, "with little flexibility in the face of climate change." As famed paleontologist Richard Leakey said at the 2005 Stony Brook forum on protected areas, they are "not protected against the ravages of climate change."[40]

The list of famed parks and World Heritage Sites under great pressure from human development, and now also affected by climate change, reads like an eco-tourist's dream itinerary: the Everglades and West Bengal's Sundarbans mangrove forest, the Great Barrier Reef and Florida Keys, the Monteverde cloud forest of Costa Rica and the Daintree rainforest of northern Queensland, Glacier National Park and Mount Kilimanjaro, Nepal's Sagarmatha National Park, the Farallon Islands, Alaska's Arctic National Wildlife Refuge, and the Antarctic Peninsula, to name just a few. Even natural paradises that have just been discovered and are free of direct human damage are menaced by climate change. Just weeks after expeditions from Kew Gardens in London and Conservation International announced discovery of a new genus of palm tree and previously unknown species of insects, birds, frogs, and a marsupial tree kangaroo in the highlands of New Guinea, another researcher said weather records showed the place was warming twenty times faster than previously known. We may think that designating such natural marvels as protected areas is enough. But in reality, these areas' very isolation makes them highly vulnerable. As Thomas Lovejoy put it at the protected area forum, "We have to stop thinking we can protect a few postage stamps with fences around them and use up all the rest." The lands set aside in national parks and reserves, along with their ecosystem services to us, will deteriorate without strong

A rainforest canopy near the Amazon River in Peru, not far from the Brazilian border. Like all tropical forests, rainforests are a font of biodiversity and human cultures and an important element in the water cycle. Once covering about 12 percent of the Earth's land surface but now reduced by more than half, rainforests harbor about half of the Earth's species and play a crucial role in regulating weather and atmosphere. (JULY 1999)

interconnections with the surrounding land, water, and people who care about them. The reverse is likely true as well: the surrounding land, water, and people will deteriorate if the protected areas are lost. We need to protect biodiversity and whole ecosystems not for their sake alone, but also to help *us* survive climate change.[41]

THUS WE ARRIVE full circle back on that wind-blasted ridge in Monteverde, Costa Rica, searching for the lost golden toad in its tiny habitat. Is this really the first species to go extinct due to recent rapid climate change? If so, and if some of its amphibian relations have likewise followed it into oblivion, it sends a chilling message. Monteverde does indeed protect species-rich ecosystems. Besides the primary highland amphibian fauna, it harbors the greatest bird diversity in Costa Rica, more than 40 percent of Central American mammal species, 250 species of butterflies and many thousands of other insects, and more than 3,000 species of plants, including over 500 orchids. Yet it is a small park surrounded by logged forests and ranches and roads that have brought urban problems to its gates. Now it is beginning to lose species due to regional or planetary warming. It is, like most protected areas, a tiny ark carrying untold natural riches, and it is unclear where it would find safer ground. There are not enough parks and reserves to protect the species we know we have against the threats we can already expect—never mind the species yet to be studied and the threats we are just coming to understand. New dangers are building beyond the borders of individual parks, nations, and continents, encompassing the entire Earth and threatening to change the conditions that have made human civilization possible for ten thousand years.[42]

4

TOMORROW'S CLIMATE TODAY

The minute I descended from the prop aircraft in Tuvalu in the South Pacific and walked through the tiny air terminal with its throng of relatives welcoming arriving passengers, taxi drivers, and vendors, I was strongly reminded of villages in the Arctic. If not for the tropical heat and very different clothing, I could have been in Pangnirtung on Baffin Island or Nome, Alaska. The feeling of connection grew stronger as I learned more about Tuvaluans' relationship with their environment. Like the Inuit and Iñupiaq, who live off the land and ice, they have a long history of living off the land and ocean. All these societies now use modern energy sources and technology, though to a much lesser extent than Western nations. Yet they are among the first to feel directly the effects of climate change and give warning to the rest of the world.

In some very serious respects, our planetary common rights—clean air and water, sufficient food, a safe and productive existence—are threatened by climate change. The primary function of the atmosphere and ocean in this commons is to moderate temperature and the composition of the air we breathe. The atmosphere blocks harmful radiation but holds some of the sun's heat near the Earth, allowing it to harbor life; the ocean exchanges huge amounts of carbon, oxygen, and heat. The interaction creates the water cycle and weather and makes possible the habitats for all living things. It has become common knowledge that due to human-made greenhouse gases, the Earth's atmosphere near the surface warmed by a bit over 1°F (0.6°C) during the twentieth century. But now that rate of warming has tripled. Scientists know that if only natural cycles and solar radiation were at work now, the Earth would be significantly cooler than it is. The atmosphere is a very thin and fragile layer that Mario Molina, Nobel Prize winner for his research on ozone, likens to "the skin of an apple." This is the air we breathe, the source of our weather, our protection from the blasting radiation of space—and now the recipient of our excess heat-trapping gases such as carbon dioxide and methane.[1]

The air, however, has absorbed only a small amount of heat compared to the ocean. Since the 1950s, 84 percent of the total heating of the Earth, instead of staying in the atmosphere, has been absorbed by and now deeply affects the sea. This huge amount of water—71 percent of our planet's surface, with an average depth of 12,000 feet (3700 m)—resists change. Yet an analysis of two million underwater temperature readings going back to the 1950s shows a worldwide increase in ocean temperatures down to 9,800 feet (3000 m). The ocean also has taken up half of the carbon dioxide generated by human activities—which is good, because that keeps the CO_2 from creating more atmospheric warming. But in doing so the sea is becoming more acidic. The increased acidity and warmth reduce its capacity to absorb carbon emissions and threaten the very basis of the ocean food chain. As the ocean water warms, it expands to rise higher along the world's million miles (1.6 million km) of shoreline. Sea level has risen 1.5 inches (40 mm) in the past fifteen years, almost double the average rate for the twentieth century. And as we saw in chapter 1, the ongoing melting of the world's glaciers threatens to produce even higher

PREVIOUS PAGE Kids hang out on their *kaupapa,* the outdoor sleeping platform favored by Tuvaluan families, as very high tides inundate their neighborhood on Funafuti Island, 9 February 2005. Thousands of homes along the lagoon and in lower-lying areas inland are flooded during increasing "king" tides as seawater comes up through the porous coral rock of the atoll. This salt water also soaks agricultural fields.

CLIMATE CHANGE AND WATER

PETER H. GLEICK

The Pacific Institute

Water shortage problems are not restricted to the developing world. As global warming worsens, we will inevitably see changes in all the planet's water resources, which in turn will affect water availability and water quality everywhere. Indeed, such changes are already appearing.

The global cycles of climate and water overlap intimately. Water evaporates to form clouds and falls back to earth as rain and snow. Plants draw on soil moisture and return moisture to the atmosphere. Depending on rainfall levels, excess water from mountain snowmelt flows into groundwater aquifers or streams and rivers. These are all part of our natural climate system—and all are vulnerable to human meddling.

Over the past century, humans have built a complex infrastructure to provide clean water for drinking and industrial use, dispose of wastes, facilitate transportation, generate electricity, irrigate crops, and reduce the risks of floods and droughts. This infrastructure has brought great benefits to society, but with substantial economic, social, and environmental costs. The average person takes for granted the dams, aqueducts, reservoirs, treatment plants, and pipes that insulate us from a naturally variable climate. Indeed, the water systems we have built permit us to forget about our dependence on climate.

Almost.

Increasingly compelling scientific evidence suggests that humans are changing the climate, and with it the reliability and quality of our water supply. Complex impacts now seem unavoidable. Although most water managers are trained to assume that the future is going to look like the past, the message from climate science over the past few decades is that that assumption is no longer true. Our children's climate will differ from our own.

The details of future climate change depend on the nature and intensity and climatic effect of greenhouse gas emissions. Many factors are difficult or impossible to forecast, including decisions of governments and individuals, deployment of alternative energy systems, and population sizes and affluence. Even if these factors were all well understood, there would remain important unresolved questions about how the climate will respond to greenhouse gases. Many crucial aspects of the hydrologic, or water, cycle are imperfectly understood.

Yet not everything is uncertain. We have learned much about the vulnerability and sensitivity of water systems, and we are exploring technologies and policies that will help us cope with adverse impacts of climate changes, and take advantage of possible beneficial effects.

What can we, potentially, expect for the planet's water cycle? We may see significant changes in the timing and amount of runoff in our rivers as rainfall patterns change, as rising temperatures affect snowfall and snowmelt, and as storms alter in intensity or frequency. Ironically, while many water managers worry about prolonged dry spells, the Intergovernmental Panel on Climate Change has warned that "the flood related consequences of climate change may be as serious and widely distributed as the adverse impacts of droughts." This group raised concerns about possible dam and levee failures in advance of the disasters caused by the 2005 hurricane season.

Higher sea levels also appear inevitable, which will affect water quality in coastal marshes and aquifers. In addition, not only will water temperatures and runoff flows, rates, and timing fluctuate more, but the ability of watersheds to assimilate wastes and pollutants will likely be compromised as well. Serious concerns for ecosystems include changes in vegetation patterns, possible extinction of endemic fish species, declines in wetlands and waterfowl populations, and degraded stream health.

Climate change will affect all aspects of water demand and use. Nearly 80 percent of the water used by humans goes to irrigated agriculture, vital for the global food supply. Although only 18 percent of the planet's cropland is irrigated, those lands produce over 40 percent of all our food. As the climate changes, so will the challenge of feeding the world's burgeoning population. Another vital challenge that must take the potential impacts of climate change into consideration involves providing the billions of people who lack basic water services with adequate safe water and sanitation, as specified in the UN Millennium Development Goals.

The availability and quality of water is directly linked to human health. Changes in climate will affect the viability of disease vectors like mosquitoes that carry malaria or dengue fever. The distribution of cholera is affected by climate, including El Niño frequency and intensity, temperature, and ocean salinity.

Finally, the impacts of climate change on water resources have the potential to affect international relations, where shared watersheds can lead to local and international political disputes. Humans have a long history of battling over water resources, and changes in climate may make these disputes more likely.

As the climate changes, we have no choice but to try to adapt by changing how we manage water systems. We must move to a "soft path" for water, which integrates the infrastructure of dams, aqueducts, and centralized treatment and distribution centers with flexible policies of conservation and efficiency, smart economics, public participation and accountability, and new forms of storage and supply. Prudent planning also demands that we maintain strong climate and water research programs to address uncertainties and unknowns. Water managers and policymakers must start considering climate change as a factor in all decisions about water investments and the operation of existing facilities. How will climate change affect water resources? Once we've thoroughly analyzed that question—and now is the time—we can begin planning how best to adapt to those changes that cannot be avoided. ✦

levels. This chapter explores the unstabilizing and sometimes downright dangerous impacts of sea level rise and climate change on our homes, our health, and our food and water.[2]

FROM PARADISE TO POLDER, THE WATER IS RISING

Life is getting riskier for people who live at the narrow boundary between land and sea. That is why I went to Tuvalu. Tuvalu has no industry, burns little petroleum, and creates only about 5,000 tons of CO_2 emissions a year, compared to the 6 trillion tons produced by the United States. The 11,800 Tuvaluans live on nine coral atolls, with a total land area of 10 square miles (25 sq km), scattered over 500,000 square miles (1.3 million sq km) of ocean south of the equator and west of the International Dateline. Here it is already tomorrow in more ways than just the time zone: Tuvaluans face the possibility of being among the first "climate refugees." The growing intensity of tropical weather, the increase in ocean temperatures, and rising sea level—all documented results of a warming atmosphere—are making trouble for this island nation. Poni Faavae, an environmental official, told me, "We are going to need to move. We need to go somewhere." Former assistant environmental minister and now assistant secretary for foreign affairs Paani Laupepa added, "Our whole culture will have to be transplanted."[3]

Tuvalu's highest elevation is 15 feet (4.6 m), but most of it is no more than 3 feet (0.9 m) above the sea. Several times each year the regular lunar cycle of tides, riding on the ever higher mean sea level, brings the Pacific sloshing onto roads and into neighborhoods. Along the World War II–vintage runway on the main island of Funafuti, seawater bubbles right out of the coral, forming puddles that eventually cover part of the tarmac. Although long experience with tropical insects and cyclones has led most Tuvaluans to perch their houses on short pilings, nowadays the recurring high tides regularly push water up into neighborhoods where not every home is elevated. In February 2005 the tides were driven especially hard against the lagoon shore of Funafuti by unusual westerly winds from a tropical convergence zone, resulting in severe erosion. The island's single main asphalt road is only about 6 miles (10 km) long, yet where it runs along the lagoon, seawater and coral rocks thrown up by the tide blocked traffic. Hundreds of wood-frame and corrugated metal-roofed homes, as well as several churches, built right on the lagoon, were drenched by the wind-driven waves riding on the higher tides. Fishing, the main means of procuring food, was dangerous, if not impossible, during this time.

nearly 16 million living within a meter of high tide, climate change and sea level rise have become crucial issues. One symptom that residents notice is riverbank erosion, which, the country's former national meteorological director, M. H. Khan Chowdhury, told me, "takes away about 8,700 hectares [19,000 acres] of land every year. About one million people are directly or indirectly affected" by this each year in Bangladesh. Dr. Chowdhury belongs to a group of retired government meteorologists and geologists who believe that "people are not very much aware of the effect on them of sea level rise" in a warming climate. And indeed, effects of climate change have not yet received extensive scientific study in Bangladesh. One review published by the Organisation for Economic Co-operation and Development (OECD) identified several threats, including heavier river flows as Himalayan glaciers melt, more intense cyclones and rainstorms, and the inexorable rise of the ocean. These are deadly in combination, as when, in 1998, heavy snows and rains in India and Nepal combined with increased monsoon downpours and high tides to flood out more than twenty million Bangladeshis. Hundreds were killed.[10]

On a boat journey south of Dhaka in June 2005, I witnessed how villagers living on the coastal plains of Bangladesh contended with shifting fresh-water rivers and tidal flow from the Bay of Bengal. The banks along the river channels were seldom more than 3 feet (1 m) high, and usually much less. Normal high tides easily overtop these mud banks and flood into planted rice fields. Residents of two villages being eroded told me of faster riverbank loss and of being unable to predict the river's rise in time to move houses and rescue families. In the tiny village of Char Kalmi on the western shore of Bhola Island, agitated townspeople pointed at the spot in the river where, in April 2005, the mosque was precipitously washed away. This happened not at the height of the monsoon season, when flooding is common, but at the end of the dry season, when most water-level changes are from the tides. Although a dike protects part of the island, many people live outside of it. In some villages, boat landings are constantly being undermined, and some villages become islands during high tides. In the main town of Bhola, the weather office had to be moved away from encroaching waters. Some villages have migrated more than 4 miles (7 km) since the 1980s, as residents leapfrog their houses away from the river's edge.[11]

Recent measurements show that sea level was rising more than 0.16 inch (4 mm) per year in the Bay of Bengal—a figure that is above the world average now. A Bangladesh government report warned of the loss of Bengal tigers in the Sundarbans, the world's largest mangrove forest and a World Heritage site, and threats to hundreds of species of birds and other animals as salt water invades. About six million people would be displaced in Bangladesh by a 20-inch (50 cm) rise, according to the IPCC, and, just as dire, rice production would be flooded out in the most productive part of the country.[12]

Bangladeshis are among the world's poorest people, with an average annual income of only about $440 (1 percent of U.S. per capita income). Not surprisingly, their contribution to global warming is also small, compared to that of residents of developed nations. The average CO_2 output of eighty Bangladeshis is about equal to the output of one American. Despite their poverty, Bangladeshis have much in common with the prosperous Dutch. People living behind the dikes on the Rhine-Meuse Delta are facing a threat from melting in the Alps and rising seas. The modern diking of Holland, one quarter of which lies below sea level, dates from a great storm that struck January 31–February 1, 1953. The sea poured into polders and unprotected land across the country, killing 1,835. Today the sixteen million Dutch live, and

70 percent of the gross domestic product is created, behind 2,300 miles (3700 km) of primary dikes. (Thirty percent of the nation is undiked.) The Netherlands is reacting to the mounting threats associated with global warming not only by strengthening these barriers but also by planning to allow some flooding, which will allow wetlands to become reestablished or expand in the delta as a buffer zone.[13]

So, too, should Italians feel at one with distant Pacific Islanders. Tidal flooding is a common event in Venice, the European symbol of rising seas and an illustration of how difficult it is to prevent damage to irreplaceable coastal cities. Venice has been sinking for hundreds of years, perched as it is on unstable sediments of the lagoon. Floods from upstream are a big factor. But the rising Adriatic Sea is rapidly exacerbating the problem. From combined effects, the sea has come up about 20 inches (50 cm) since the eighteenth century, and there are more than five times as many flooding events now as during the 1920s and 1930s. At the current rate the sea will rise another foot (30 cm) in this century. Some winters, the *acqua alta* covers the Piazza San Marco, only 2 inches (5 cm) above the high-tide level, more than one hundred times. Italian officials have decided to construct elaborate tide dams at lagoon entrances. Environmental groups and some scientists warn that higher tide and storm levels will soon overcome these defenses, and the dams may isolate the lagoon from the natural flushing it needs to remain a viable ecosystem.[14]

In Britain, they've already built a tide gate on the Thames, just downstream from the center of London. Looking like a bridge designed by Frank Gehry, with great sculpted metal housings for the gateworks, it was closed against extreme high tides only once in the first few years after its construction in 1987. By the late 1990s it was being raised more often, and now, according to former British science adviser Sir David King, flooding tides force its use six or seven times a year. Additional issues plague London as well as Holland, Venice, and other coastal locations, including land subsidence, groundwater pumping, and overpopulation. These make more urgent the extra threat on shorelines from a rising, warming sea. Sea level may get higher this century than the IPCC report indicates, owing to faster-moving glaciers (mentioned in chapter 1). The rate of change has been keeping pace with temperature rise, and that, too, points to as much as 3 feet (1 m) more sea by 2100. Coastal and estuarine landscapes— barrier islands, reefs, seagrass beds, tidal flats, dunes—and artificial protective structures are less likely to survive or provide protection for inland areas when storms ride in on a higher ocean.[15]

Hurricane Katrina brought this threat home with frightening power to the subsiding Mississippi Delta, Gulf Coast, and New Orleans. The entire United States watched in disbelief as years of predictions by journalists and scientists—physical, social, and political— came horribly true. The growing strength of storms, the inadequacy of levees, the limitations of industrial dredging and channeling as flood protection, the vulnerability of four thousand oil and gas platforms producing a third of the U.S. daily oil supply, the lack of storm buffering due to widespread removal of wetlands, the difficulty of evacuating the ever-increasing coastal population—most of this tragedy was foretold. There were huge losses to commercial buildings, homes, the transportation infrastructure, the fishing industry, tourism, crops, and livestock, in urban and suburban as well as rural areas. Business, utilities, the fuel supply, and communications were interrupted. A diaspora of a million souls fled the Gulf Coast, eventually reaching every state of the nation; more than a quarter of them have not returned. Survivors suffered

Acqua alta, or tidal flooding, of the Piazza San Marco is a common event in Venice. The city has been sinking for hundreds of years, and floods from upstream are part of the problem. But the rising Adriatic Sea is rapidly exacerbating the subsidence. The sea has come up about 20 inches since the eighteenth century, and at the current rate it will rise a foot in this century. (1984)

for lack of food, shelter, health care, and transportation. Every effect was heightened by social inequality and the failure of governments to function. Unfortunately, Katrina will not be the last large storm to wreak havoc on a major urban area. In 2006, hurricane scientist Stephen Leatherman singled out New Orleans, the Florida Keys, Miami, Galveston/Houston, eastern Long Island, the Tampa Bay area and Wilmington, North Carolina, as most vulnerable to extreme storms. Lake Okeechobee, Florida, the Mississippi coast, and Cape Hatteras were also high on his list. As a report from Harvard University's Center for Health and the Global Environment put it, "Climate change may not be a threat to the survival of our species, but it is a threat to cultures, civilizations and economies that adapt to a particular climate at a particular time."[16]

One factor raising the risk to American coastal cities is the number of tropical storms in the Atlantic. In the 1970s and 1980s climatic oscillation reduced the number of hurricanes, but since about 1995 the climate has been in a positive phase, and, according to Kevin Trenberth, head of the Climate Analysis Section of the National Center for Atmospheric Research (NCAR), the number of storms has increased to a level that is considered near normal. This fifty- to seventy-five-year cycle is apparently not affected directly by global warming, nor is the total number of tropical storms each year across the globe, which remains constant at about eighty-five. No scientist could honestly blame Katrina, or Rita, or any other single storm on global warming. Storm intensity, however, is another matter.[17]

Warmer seas and atmosphere mean that, once begun, hurricanes are likely to get stronger or develop faster. According to NOAA's National Climatic Data Center (NCDC), "Recent studies . . . indicate that . . . the destructive power of hurricanes has generally increased since the mid-1970s, when the period of the most rapid increase in global ocean and land temperatures began." One of these studies shows that the power of hurricanes worldwide was up about 50 percent over thirty years; another reports a near-doubling of category 4 and 5 storms. The disastrous North Atlantic hurricane season of

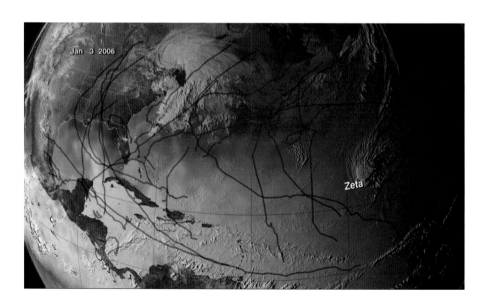

Satellite imagery of the Atlantic Ocean and Tropical Storm Zeta, the twenty-eighth storm of the 2005 season, which actually spun on into 2006. Storm tracks of all the other tropical storms and hurricanes during this record season are shown on this image. Increased ocean temperatures are linked to greater storm strength, and around the world, not just in the Atlantic, hurricanes appear to be growing more powerful. (NASA)

2005 appeared to confirm this research. An unprecedented twenty-eight tropical storms formed, according to NOAA, of which fourteen became hurricanes. For the first time in a single season, three storms, Katrina, Rita, and Wilma, strengthened to category 5 in the Atlantic Basin, and two others went to category 4. Katrina progressed from category 1 as it crossed Florida to category 5 when it moved out over the exceptionally warm Gulf of Mexico, taking two days to build strength (it hit New Orleans as a strong 3). Sea surface temperatures "were at record high levels in the subtropical Atlantic critical region for hurricanes," said Trenberth. "That warming is linked to global warming." MIT meteorologist Kerry Emanuel wrote in *Nature* that, looking beyond the 12 percent of hurricanes that occur in the Atlantic Basin, around the world "levels of tropical storminess are unprecedented in the historical record." However, the hypothesis that global warming directly intensifies storms has been thrown into doubt by new research. One predictive study led by Emanuel himself did show storms getting stronger, but results varied so much that the greenhouse gas correlation was uncertain.[18]

That there is a connection between greenhouse gases, increasing heat energy, and tropical storm intensity seems clear, but not all climatologists consider it confirmed. Colorado State University emeritus climatologist William Gray, Christopher Landsea of NOAA's National Hurricane Center, and University of Virginia climatologist Patrick Michaels offer alternative analyses. They generally do not deny that surface ocean temperature has a strong effect; Gray cites it as a major factor when he predicts coming hurricane seasons. But they argue that the detailed data record is short, extending back only about sixty-five years, and that there are different ways to parse the factors in that record. These scientists often point to wind shear, circulation, strong natural climate cycles, and local effects within ocean basins as overriding the effect of sea surface heat. To be sure, the increase in this heat has not been uniform, yet it is hard not to be struck by the graphs produced by Emanuel and others showing hurricane strength and sea surface temperature rising in an undulating curve that echoes the rise in air temperature. The 2006 Atlantic season was below normal—a fact ascribed to higher wind shear and much lower South Atlantic sea temperatures, perhaps due to an oncoming El Niño. Other oceans, however, generated huge typhoons and cyclones that killed hundreds in China and the Philippines, and also hit Australia and Japan. The World Meteorological Organization stated in late 2006 that the research is not yet "definitive" on the connection between greenhouse gases and storm intensity, but most research shows some connection with sea surface temperature. At least one study hypothesized that it was temperature differences across oceans that caused more storms in the late 20th century.[19]

Another way of measuring the impact of hurricanes and rising sea levels is to look at the physical and monetary damage they inflict. The United States is heavily populated along its coasts, with more than 3,500 people moving there every day. Although "coastal counties constitute only 17 percent of the total land area" of the United States excluding Alaska, according to NOAA, they "account for 53 percent of the total population"—153 million people. All but two of the twenty-five most densely populated counties in the country are along the coast. Florida had the greatest percent population change between 1980 and 2003, up nearly 75 percent, and most of the newcomers have moved within a few miles of the ocean. The apartment complexes, homes, and businesses that crowd the Atlantic shoreline are literally and metaphorically on the edge. After four hurricanes in 2004 and three in 2005, and with a generally more active Atlantic storm regime in play, it is not unreasonable to say that the entire coast of Florida is in the meteorological crosshairs, with more financial disasters and loss of life in store.[20]

Beaches are being washed away by high tides, millions of dollars are being spent on beach armoring and sand replenishment, and ever higher sea levels only raise the threat from storms. Rising sea level is also inundating mangrove swamps, threatening low-lying islands, and driving seawater into Everglades National Park, a World Heritage Area. These developments cast a shadow over multibillion-dollar efforts to restore the northern part of the park after years of canal digging and water overuse by farming and suburbia. To the north, on Cape Hatteras in North Carolina, the U.S. National Park Service recognized the futility of trying to protect America's most famous lighthouse from the eroding shoreline. In 1999 it moved Hatteras Light back 2,800 feet (850 m) from the surf. Erosion along the cape has been proceeding at a pace of about 12 feet (3.6 m) per year recently, leaving house after house stranded in the surf, awaiting destruction by the sea.[21]

This situation cannot improve. According to scientists at the Laboratory of Coastal Research at Florida International University, the rate at which sandy beaches erode is about one hundred times the rate of rise in sea level. As part of this process, narrow barrier islands migrate landward as storms wash more and more sand toward the lagoon. This of course means that more and more houses, roads, and infrastructure along populated spits like the Outer Banks are going to fall. Federal insurance usually guarantees money for rebuilding after storms, so local officials continue to bulldoze sand back onto beaches and reestablish roads, only to have events repeated with sea level a bit higher. Like King Canute, they cannot command the sea.[22]

Sea level rise is fast resculpting the geography and ecology of America's great bays. Habitat and feeding areas for the huge migrations of breeding birds from the Arctic are being pushed farther and farther up the beaches. On Delaware Bay in April, I saw the beach narrow to a strip right below corrugated storm drain pipes, while thousands of red knots, turnstones, and sandpipers were fueling up on horseshoe crab eggs. Ornithologist Hector Galbraith and his associates say that "over the next few decades global warming is predicted to result in an acceleration in the current rate of sea-level rise, potentially inundating many low-lying estuarine areas and intertidal habitats"—including Delaware, San Francisco, Humboldt, and Willapa Bays.[23]

Even more drastic changes are occurring in Chesapeake Bay, where combined sea level rise and subsidence is more than twice the world twentieth-century average: about one foot in one hundred years at Baltimore and 1.3 feet at the mouth of the bay. Crucial islands such as Poplar and Smith and wetlands such as the Blackwater National Wildlife Refuge are succumbing rapidly to the encroaching waters. At Blackwater, nearly half the original 17,000-acre (6900 Ha) reserve has flooded, and pine forests used by bald eagles are dying as salt water invades their roots. In the middle of the Chesapeake, at least thirteen islands have disappeared, including Cockey's, Herring, Powell, Punch, and Sharps, which played roles in the early exploration of the eastern seaboard. More than 70 percent of marshland in the Delaware and Chesapeake Bay estuaries is degraded, affected by higher tides.[24]

A house on Cape Hatteras, North Carolina, undermined by increasing erosion to the dunes between 1999 (above) and 2004 (below). Even without strong hurricanes the cape is losing 12 feet a year, leaving house after house stranded in the surf, awaiting destruction.

On Delaware Bay in April a normal high tide narrowed the beach to a strip right below storm sewer outfalls during the annual migration of thousands of red knots, turnstones, and sandpipers. Chesapeake, San Francisco, Humboldt, and Willapa Bays, along with river deltas and smaller estuaries, are also experiencing an accelerating rate of sea level rise that is inundating many islands and intertidal habitats. [APRIL 2001]

Everyone notices: greater changes and extremes in weather seem more and more common lately. Increasingly, news stories confirm many people's observations that spring is coming earlier, rainstorms are pelting down harder, snowfall is less predictable, heat waves are more common, and droughts afflict larger areas. It is true that around the world many (though not all) long-term observations and scientific studies corroborate these impressions. Do these apparent trends constitute convincing evidence that global warming is making all types of weather more severe?

To be sure, every year has its incredible weather statistics, and some past events have been far worse than recent events. Even though a current dry spell in the American Midwest and West is expanding into a sixth year and each summer hundreds of cities set heat records, the "dust bowl" drought of the 1930s was drier and affected a greater area of the United States. The 1988 drought is second only to Katrina as the most expensive American weather disaster. (However, population and water use in the West have increased so much that today's drought may affect more people.) Twentieth-century droughts are in turn surpassed by the decades-long droughts of the mid-sixteenth and late thirteenth centuries (the earlier one possibly contributing to the disappearance of the Anazasi). Because daily and seasonal weather can fluctuate widely in the normal course of events, climate analysts go to great pains to use solid, peer-reviewed statistical methods to smooth out any short-term changes. Climate, after all, is the long-term record of weather.[25]

As noted earlier, one indicator of weather severity is the amount of monetary damage brought about by storms, floods, and droughts. Of the sixty-seven American weather-related disasters exceeding one billion dollars' damage between 1980 and 2005, 87 percent occurred after 1988 (64 percent of the time period). Such an accounting is of great importance to the insurance industry, governments, and property owners (though, perversely, it can lead to extreme weather events being considered, at least in part, as an economic stimulus, since money must be spent to clear debris and rebuild). However, considering that property values are increasing and more and more people live in danger zones along coasts and in floodplains, the amount of monetary damage is not a precise measure of weather severity.[26]

Insurance companies nonetheless are well aware of the connection between climate and extreme weather events and are among the most vocal in urging action to reduce greenhouse gases. Munich Re, which underwrites insurers, reports on its website that the frequency of "great natural catastrophes" more than doubled between 1960 and 2005, with the increase due to more weather disasters. Harvard's "Climate Change Futures" report shows that there were four times more weather-related disasters in the 1990s, including wind, flood, drought, and epidemic attacks, than in 1950–59. Swiss Re, another reinsurer, commented in its 2004 annual report that that year had been an especially severe one for storms: "Climatologists attribute the high windstorm frequency to above-average sea-surface temperatures and the higher year-round average temperatures measured in the last decade." Munich Re, meanwhile, warned that "flood risk will increase as precipitation falls more frequently and over larger areas in the form of rain rather than snow. . . . Also, the rise in sea levels that is to be expected as a result of global warming will aggravate the risk of storm surges and beach erosion on all coasts of the

Flooding in Gurnee, Illinois, in May 2004 after record rains in Wisconsin. Globally there has been a 1 to 2 percent increase in precipitation per degree of average temperature rise, but how and where this water falls is very uneven. Rainfall increased 7 percent over the lower forty-eight United States since 1895, and half of the wettest years since 1950 have occurred just since 1990. Deluges are increasing in intensity even where the total rainfall has remained constant.

world." Insurance firms are concerned, of course, with their liabilities and profits, but governments and property owners should also pay attention to uninsured losses, which according to some statistics are now double the value of insured losses and rising faster than economic growth. In flood- and hurricane-prone locations, insurance is becoming increasingly hard to buy. [27]

FROM ANALYZING the weather record itself, scientists can say with 90 percent certainty that daily high temperatures are creeping up and that winters and nights are warming. As we saw in chapter 3, there are fewer days with frost, and the growing season is longer by eighteen days in Eurasia and twelve days in North America relative to thirty years ago. In Maine, Wisconsin, Alaska, and elsewhere, rising temperatures mean thinner ice on lakes in winter; lakes freeze almost six days later now and melt out more than six days earlier than 150 years ago. According to eighty-four years' worth of records on the Nenana River near Fairbanks, Alaska, spring breakup (known to the minute because Alaskans place bets on it) has advanced by five and a half days. Wasilla, near Anchorage, no longer serves as the start of the yearly Iditarod sled dog race to Nome because good snow conditions can only be found 30 miles (50 km) or more farther north. Field station reports and satellite images show that snow cover is decreasing throughout the Northern Hemisphere. Together with the advancement of spring warmth, this means that peak stream flow occurs about nine days earlier across the American West. [28]

One of the strongest forces in our atmosphere is water vapor. The warmer the air becomes, the more water it can hold—a process that increases global warming because water vapor is itself a greenhouse gas. Although not as powerful per molecule as carbon dioxide or methane and very short-lived in the air, water vapor, because it is so abundant, actually has up to twice the real effect of the other gases. Heat and water vapor are the raw materials in storms. Thomas Karl, director of NOAA's National Climatic Data Center, and NCAR's Kevin Trenberth wrote in *Science* recently that "basic theory . . . [and] evidence all confirm that warmer climates, owing to increased water vapor, lead to more intense precipitation events even when the total precipitation remains constant, and with prospects for even stronger events when precipitation amounts increase." Globally, there has been a 1 to 2 percent increase in precipitation per degree of average temperature rise, but how and where this water falls is very uneven. Rainfall increased 7 percent over the lower forty-eight United States since 1895, and half of the wettest years since 1950 have occurred just since 1990. The amount of rainfall coming in extreme events rose by 14 percent (mostly in the past thirty years). Across thousands of weather stations from Canada to South Africa and Asia, very heavy precipitation events have been more common in the past thirty to forty years. Extreme monsoon deluges in India have more than doubled compared to fifty years ago, even though the total amount of rainfall has stayed about the same. This statistic is background to the incredible 37 inches (94 cm) that swamped Mumbai on 26 July 2005. Analysis by the Hadley Centre for Climate Prediction and Research of the British Met Office confirms that river flows are increasing, not only because of changes in weather, but also because of the effect on plants of elevated CO_2, which slows the drawing up and release of water through their leaves. The number of floods, the most common natural disaster, increased in Europe from 1975 to 2001.[29]

Looking to the other extreme, a 2004 study led by Aiguo Dai of NCAR found that the area of the

Earth suffering from drought had more than doubled in the past thirty years, with about 30 percent of the planet now characterizable as very dry. Drought slows plant uptake of CO_2, creating a vicious cycle that leads to higher levels of greenhouse gases. At present, three billion tons of soil and grit are being blown around the planet, let loose from drought areas as well as from agricultural lands and deserts. The effects worldwide are severe, ranging from choking sandstorms in Beijing and coughing and irritated eyes among Canadians to the smothering of coral in the Caribbean and the darkening of Greenland's glaciers, which in turn causes more rapid melting. According to Dai, "Droughts and floods are extreme weather events that are likely to change more rapidly than the average climate." The study reported that "global warming not only raises temperatures, but also enhances drying. . . . The increased risk of drought duration, severity, and extent is a direct consequence." In other words, despite imperfect or no data for some parts of the world and other places that show no significant change, the overall trend seen in drought and rainfall confirms the intensification of the water cycle that was predicted to occur as the atmosphere warmed.[30]

Another complex influence on climate playing out in the atmosphere came to light when commercial flights were grounded after the attacks of September 11, 2001, and jet condensation trails (contrails) disappeared. Records from four thousand weather stations across the United States revealed an increase in the range of daily temperatures for those days: it got warmer by about a degree during the day, or colder at night, or both. When air traffic resumed on September 15, the temperatures returned to pre−9/11 ranges.[31]

It may not be too surprising that contrails, like regular clouds, impede some daytime radiation and also keep some warmth in at night. But that four-day window of measurements helped focus attention on a bigger change that had been observed worldwide: "global dimming." Since about 1950, industrial pollution—tiny particles of soot, dust, and chemicals—has been filtering out about 4 to 5 percent of all sunlight headed toward the ground. These particles, called aerosols, block or reflect sunlight directly and also create clouds (having somewhat the same effect as contrails). The impact of these aerosols since the 1950s has been dramatic: "In the United States," according to meteorologist Beate Liepert of Columbia University's Lamont-Doherty Earth Observatory, "sunlight decreased 10 percent. . . . In Hong Kong, the figure was a shocking 37 percent." Other atmospheric scientists have noted plumes of this pollution flowing from industrial areas and creating new weather patterns. The biggest issue with global dimming, though, is that it actually holds back some global warming: that is, the Earth would be getting even warmer without it. It is an issue because since the 1990s, as air quality has improved, the dimming has decreased, meaning that temperatures will likely go up faster now. Climate scientists have included more and more details about aerosols and clouds in their predictive computer programs to try to figure out how much. It's ironic that by cleaning up pollution, we may be revealing the depth of global warming. This underscores the fact that, just as we saw the danger in local industrial and automobile pollution and took steps to control it, so we need to see the connection between carbon dioxide and other greenhouse gases and the warming that jeopardizes so much of the world.[32]

Lake Powell, Utah, from the air, looking over Dangling Rope Marina, September 2004. The lake was only about one-third full, leaving behind a wide "bathtub ring" in the sloping canyon. About one-third of the United States is experiencing drought, especially the Southwest, and the National Center for Atmospheric Research found that the area of the Earth suffering from drought more than doubled in the past thirty years.

As severe weather events reach a crescendo, the lives of a growing number of people are torn apart. The Katrina disaster, with over 1,300 people killed, a million evacuated, and hundreds of thousands scattered across the Union, showed us how wind, rain, and flooding leave in their wake not only death but also injured and homeless people more susceptible to disease. "There is now recognition that global climate change poses risks to human population health," according to the World Health Organization (WHO). Based on 2000 figures for mortality from floods, diarrhea, malaria, and malnutrition, WHO conservatively estimates that some 150,000 deaths each year are caused by global warming. (Although regional correlations are strong, no direct connection has yet been shown scientifically between worldwide average climate warming and specific diseases, so such approximations remain speculative.)[33]

"Adverse health conditions cluster around extreme events," wrote Paul Epstein, associate director of the Harvard Center for Health and the Global Environment, in a 2004 briefing paper. In other words, warming favors disease. Droughts in the American Southwest, for example, have aided the spread of hantavirus by disrupting the normal balance of predators and the rodents that carry the virus. And the steady increase in temperature has speeded the spread of malaria, dengue, and other insect-borne diseases of the tropics, according to Epstein. Other medical authorities question blaming temperature alone, citing forest clearance, dense population, drug resistance, migration, and political decisions that inhibit preventive measures. Nevertheless, highland areas (above a mile in elevation) in Latin America, Africa, India, and the Pacific, long free of these diseases because they were too cool for the insect carriers to survive, are now seeing more and more cases. According to a study of four areas in Africa between 5,000 and 10,000 feet (1500–3000 m), a temperature increase of less than 1°F (0.5°C) not only correlated with the reestablishment of malaria since the 1970s but also accelerated the development of the mosquitoes. Malaria, which had been mostly eliminated from northern areas, has returned to the United States, the Korean peninsula, and parts of southern Europe. Dengue fever, previously limited to about 3,200 feet (1000 m), has appeared above 5,500 feet (1700 m) in Mexico, and its mosquito vector, *Aedes aegypti*, has been seen even higher in Colombia. The West Nile virus is currently spreading across the United States and has adapted to more than 110 species of birds and many species of mosquitoes. It probably is aided by warmer average temperatures, heat waves, and heavier rains. The disease, first noted in the United States in 1999, struck 2,539 people in 2004, causing 100 deaths; those numbers increased to 4,269 cases and 177 deaths in 2006. It has spread to all but five states and even into the cooler climate of eight Canadian provinces. Ragweed pollen is increasing in step with increases in carbon dioxide, and asthma is affecting more children. A study of climate effects on health led by Jonathan Patz in 2005 concluded that temperate areas and regions of the Indian and Pacific Ocean Basins were particularly vulnerable to increasing sickness and death.[34]

Hurricane, floods, and heat waves have recently brought severe suffering to developed nations. Typically, however, adverse health effects fall disproportionately on people in Africa, the eastern Mediterranean, South America, and Southeast Asia, poor regions with fewer health and social welfare agencies. Already more than one billion people lack safe drinking water, and two and a half billion live without

DISEASE RISKS FROM CLIMATE CHANGE

PAUL R. EPSTEIN

Harvard Medical School

While no one event is diagnostic of climate change, the pattern of unusually severe weather since 2001—prolonged droughts in the United States and Europe, intense heat waves, violent windstorms, and widespread "hundred-year" floods—points to a changing climate. By the fall of 2005, what were once worst-case scenarios for extreme weather events in the United States were no longer unimaginable. Catastrophic hurricanes Katrina, Rita, and Wilma killed more than 1,400 people and displaced a million others. They spread toxins and microorganisms throughout the Gulf Coast and destroyed wetlands and barrier islands. Damage to oil production and refining facilities exposed the vulnerability of the energy sector to increasingly intense storms associated with climatic instability.

Health is a key indicator of environmental change, and a variable climate poses risks for individual, public, ecological, and economic health. Heat waves have the most immediate impacts and are projected to take an increasing toll in both developed and developing nations. The extraordinary 2003 summer heat wave in Europe caused up to thirty-five thousand deaths in five nations, extensive wildfires, and widespread crop failures. A changing climate can be seen also in more serious floods, such as those that inundated central Europe in 2002 and again in 2005.

Extreme weather events affect the transmission of infectious diseases. Heavy rains can create insect breeding sites, drive rodents from burrows, and contaminate clean water systems. Conversely, drought can spread fungal spores in dust, spark fires that lead to respiratory illness, and amplify the reach of mosquito-borne viruses, like West Nile.

Climate change influences the range of infectious diseases as well. Several key diseases or their vectors are already ascending mountain ranges in the Americas, Africa, and Asia, in the very places where glaciers are in rapid retreat, plant communities are migrating upward, and temperatures are rising more rapidly than at lower elevations.

In the marine environment, ocean warming contributes to harmful algal blooms, which can cause shellfish poisoning, provide a reservoir for cholera and other bacteria, and lead to hypoxic "dead zones," as in the Gulf of Mexico. Extreme rain events spawned by warming in the deep ocean may also flush organisms, chemicals, and nutrients into bays and estuaries, where plankton blooms are likely to occur.

Excess carbon dioxide in the atmosphere has health consequences as well. Experiments have shown that ragweed grown with elevated carbon dioxide levels produces a lot of pollen. Weedy plants can proliferate under similar conditions. Pioneering trees that spread quickly after disturbances, such as maples, pines, birches, and quaking aspen, also appear to be boosting production of seeds, cones, and pollen in response to elevated carbon dioxide levels, and some soil fungi are producing more spores. Since 1980, asthma has quadrupled in the United States and in many other regions; stimulation of plants and fungi by CO_2 may be an important factor.

Crop pests and diseases are also affected by climate change. Many of the insect herbivores and vectors of viruses, such as whiteflies, respond to warming in the same way as do insects that carry human diseases. Extreme weather events are also conducive to outbreaks of disease and infestation: flooding encourages fungal growth and nematodes, while drought promotes outbreaks of aphids, locusts, and whiteflies.

These problems are having an impact on crop yields worldwide. In Africa, for example, Ethiopia, Somalia, Sudan, Kenya, Uganda, and Tanzania all suffered in 2006 from an extended drought. Crops wilted, animals starved, and famine spread. Likely future impacts include malnutrition, which stunts the growth of children, retards their intellectual development, and increases their susceptibility to lethal infectious diseases.

Public health is central to meeting the UN Millennium Development Goals, which include eliminating extreme poverty and hunger, improving maternal health and reducing child mortality, combating the most devastating diseases, and ensuring environmental sustainability. Ultimately, human health depends on environmental integrity and social development, as well as how we power that development. Today climate influences the health of humans and the environmental systems that underlie public health. Given the proper incentives, powering development clearly can be good for public health, stimulating environmental growth and helping to stabilize the climate. ✦

adequate sanitation—a situation that will only be exacerbated by continued drought and the melting of mountain glaciers due to climate warming. Billions more depend on water- and weather-sensitive livelihoods such as fishing and farming. "The regions with the greatest burden of climate-sensitive diseases are also the regions with the lowest capacity to adapt to the new risks," Patz notes. They are also, very often, contributing little to global warming. In particular, "Africa—the continent where an estimated 90% of malaria occurs—has some of the lowest per capita emissions of the greenhouse gases that cause global warming. In this sense, global climate change not only presents new region-specific health risks, but also a global ethical challenge."[35]

Africa is of special international concern owing to the brutal intersection of human-caused disruption, disease, and climate disasters. War and brutality among tribal and political groups are constant threats in many areas. Logging and wood gathering combine with overgrazing to lay bare the soil. Seventy percent of the working people of Africa are employed on small farms, which provide most of the food, but they also are the most vulnerable to changes in the amount and timing of rainfall. Lower rainfall has been implicated in locust plagues in the Sahel, south of the Sahara, and in disruptions to the traditional planting time in Rwanda and other countries. Fourteen nations in Africa have water shortages that are linked to a half-century trend of failing rainfall. The Horn of Africa suffered under a five-year drought, only to have the rains return with such ferocity that refugee camps were washed away; hundreds of thousands were isolated by flash floods. The human suffering is made worse by a 2°F (1.2°C) rise in regional temperature over the past thirty years, culminating in the hottest temperatures on record during 2004–5.[36]

Climate scientists like NOAA's Martin Hoerling believe that many of the weather events and record high temperatures are clear fingerprints of human-made global warming. The scientists have struggled, however, to understand the complex reasons for the failure of the normal rain cycle over so many years. Apparently, the long-term shift that disrupted the rainy seasons in the Sahel and in southern Africa is being driven by changes that can't be blamed solely on higher atmospheric temperatures but rather, Hoerling wrote, on "patterns of ocean variability and/or change." The climate of Africa is partially under the influence of worldwide events like the El Niño-Southern Oscillation, for example. In the case of southern Africa, higher sea surface temperatures in the Indian Ocean can be blamed in part for the sudden and sustained aridity that began in the late 1970s and shows no sign of abating. As for the devastating dry period that hit the Sahel starting in the 1960s, Hoerling said, it appears that temperature differences between the North and South Atlantic were a main driving force—and they may have also contributed to the partial return of the rains in more recent years. "One point is virtually certain," Hoerling wrote. "Surface temperatures will continue to climb, and that alone leads to drought stress via increased evaporation, increased human and agricultural demand." But, he cautioned, predicting the future climate in Africa is still very difficult, in part because "dramatic Sahel rainfall swings . . . have historically occurred due to purely natural causes. There is no reason to believe that these natural swings will diminish."[37]

Record heat in the western Atlantic nearer the equator may be contributing to weather extremes in South America. The first hurricane ever known in the South Atlantic spun into southern Brazil in late March 2004. Coming ashore in the night, its 114-mile-per-hour (180 kph) winds hit Torres and thir-

Kenyan and Somalian refugees who lost their livestock in severe drought suddenly face unrelenting deluge at the United Nations camp near Dadaab, Kenya. More than a million people in East Africa suffered through unusually intense rains in November 2006; hundreds of thousands waited many days for food, tarps, and medical supplies to reach them as desert roads were sliced by flash floods. (PHOTO © BRENDAN BANNON/UNHCR)

teen other towns, damaging up to 90 percent of homes, and flattening corn and banana crops. As with storms in the North Atlantic, atmospheric scientists reported that the hurricane's unusual intensity was due to conditions correlated with climate change, including high sea temperatures.[38]

The worst drought in forty years gripped the Amazon in 2005. River flow gave way to expanses of mud on many tributaries, cutting river access to nine hundred towns, killing fish, and disrupting commerce. As in Africa, low water flows created conditions for poor sanitation and disease. Recent multi-year studies of how the Amazon climate operates and is changing in the face of deforestation did not predict the severity of this drought. Drying events have occurred before during El Niños, but 2005 was not an El Niño year. Analyses point to a link to record high Atlantic Ocean surface temperature, which tends to create stable, dry high-pressure zones south of the equator. However, Atlantic sea temperatures are also related to Pacific Ocean cycles like El Niño as well as a great movement of air over the equatorial Pacific called the Walker Circulation. The Walker Circulation is getting weaker, apparently as a result of climate warming. The whole tropical zone may be getting wider, bringing warmer and drier air farther north. Scientists are seeing other climate danger signs for Amazonia. Many studies document that the 1.5 million-square-mile (4 million sq km) watershed may have two equilibrium states—one that we see now in the lush, diverse, tropical forest and another that is much drier and more savanna-like. Severe drought weakens trees and invites fire, both of which push the region toward savanna conditions. The loss of Amazon forests to logging and agriculture, continuing at a rate of about 7,700 square miles (20,000 sq km), or approximately 0.5 percent, each year, is a major factor in changes to the rainfall patterns, causing heavier storms in some places and shifting the timing of rainy seasons. A change from rainforest to savanna has huge implications for the region and the planet. The Amazon basin harbors more than half the world's tropical rainforest and takes in about 8 percent of the world's CO_2 emissions. Evaporation from this vast forest drives much of the region's weather.[39]

Though the exact mechanism behind climate change in Africa and the Amazon is not completely understood, the role of global warming in the brutal European heat wave of 2003 is quite clear. The effects included severe drought, forest fires, rapid glacier melting, and prolonged high temperatures across Europe, especially in the cities. August 2003 was the hottest on record in the Northern Hemisphere. Many places had readings well over 100°F (35–40°C) day after day, and in Paris, all records for maximum and average temperature and duration of heat were exceeded during the period August 4–12. Air pollution was severe. Drought caused coolant water supplies to shrink, and power plants had to be shut down, reducing electricity for home air conditioning and fans. Attacks of heat stroke began to mount, especially among the elderly and even more sharply among those who lived in nursing homes. People died more often at night, when evening cooling did not arrive to relieve their overheating, dehydration, and respiratory difficulties. People die from these causes in any summer, of course, but in summer 2003, 14,800 more people died in France than usual, 7,000 more in Germany, 4,200 more in Spain, and 4,000 more in Italy. Across Europe, in the ten heat-struck nations—all of which have modern and efficient health care systems—the estimated death toll was at least 35,000. (The worst U.S. heat waves have killed an estimated 10,000 in a single summer.) Records showed that the 2003 heat wave was the deadliest in recorded European history. The prolonged scorching also created a perverse feedback loop, stopping plant growth in fields and forests and sending huge clouds of carbon dioxide into the air. Under normal conditions, Europe's plants take up about 137 million tons of CO_2 (125 million mt), but in the summer of 2003 they were releasing it at a rate four times as great. Unfortunately for Spain, Portugal, and parts of France, this was just part of an extended drought, the worst in the region since the 1940s.[40]

FEEDING OURSELVES

A few climatologists and botanists say that global warming will be a boon to humans because increased heat and CO_2 will spur growth of food crops. After all, they say, "CO_2 is a plant food" that helps drive photosynthesis. Indeed, most plants do grow faster with additional CO_2. Field trials are finding a downside, however. Higher CO_2 levels may reduce both the vitality and the nutritional value of crops, leading to less healthful foods and a greater potential that insect pests will eat more vegetation. These may not be problems for rich nations that can afford a variety of nutrient-rich foods as well as fertilizer and pesticides to promote plant growth and fight insects. But for poor nations that overwhelmingly rely on rice, corn, and wheat for sustenance, loss of nutritional value and plant vitality will be serious drawbacks. Grazing animals may also suffer or take longer to mature if grasses and forage provide fewer nutrients. Scientists undertaking experiments in which plants are bathed in extra carbon dioxide in the open are finding some limits to the amount of CO_2 that plants can take up and the amount of additional growth they put on. Although grassland test plots in Minnesota had a positive response to extra CO_2, a shortage of nitrogen gradually restricted their growth. These limits may apply to domestic grains, trees, and wild plants as well. If that proves to be true, it would dash hopes that plants will save the day by absorbing extra carbon dioxide far into the future. In some experiments, noxious weeds grew much faster than the preferred crop; poison ivy, for example, not only sped up 150 percent but became more toxic as well.[41]

Ozone levels also increase in industrial nations under higher air temperatures. Steve Long, who experimented with twenty-two varieties of soya at the University of Illinois, reported in 2005 that the gas, which interferes with photosynthesis and can damage leaves, cut yields by 20 percent. As far as CO_2 is concerned, he identified at least one crop pest, the Japanese beetle, that does better in high-CO_2 environments. Increasing bad weather, drought, and attacks by insects and other crop pests could cut agricultural harvests, especially of the three intensively monocultured grains, rice, corn, and wheat, that feed nearly everyone. Some of these effects have already been experienced in China.

Higher average heat is a problem too. U.S. yields of corn and soybeans have been increasing but with lesser gains in warmer years: according to one recent study, there is a 17 percent reduction in harvest per one degree increase in growing season temperature. In Philippine test fields, an increase in nighttime temperatures of almost 2°F (1.2°C) since 1979 has been linked to a 10 percent reduction in rice yields. Rice alone feeds two billion people a day, the most of any grain. Water shortages are likely to restrict some harvests as well, not to mention production of top-of-the-food-chain products like beef, which requires more than 2,500 gallons of water to produce one pound of meat. Shifting climatic zones may push optimum growing locations into areas where soil or open land are limited, not only for food crops but also for orchards and vineyards. On top of all this, food production will be affected by the increasing cost and decreasing availability of petroleum-based agricultural chemicals, farm machinery, and food transport.[42]

THE TRIPLE THREAT TO REEFS

Coral reefs constitute the most complex ecosystem on the planet, home to hundreds of thousands of species ranging from sharks to bacteria. Although they occupy a very small part of the Earth, no more than one-tenth of 1 percent, coral reefs have amazing diversity, due to infinitely variable kinds of habitat. Yet despite the immense number of species that may be present in coral reefs, "they are not wide ranging," says marine biologist Gustav Paulay of the University of Florida. "Many species are very narrowly distributed."[43]

Reefs are made of the calcium carbonate skeletons of the colonial coral polyp and as such represent a cycling and storage of CO_2, essential for maintaining the balance in the atmosphere and ocean. (Some algae secrete calcium carbonate and build another type of reef.) Reefs also offer coastal protection to up to 500 million people in eighty-six nations around the tropical zones, as well as monetary gains to these countries through fishing, recreational opportunities, and the development of new drugs from reef organisms, which together account for an estimated $375 billion in goods and services yearly, worldwide. A quarter of the fish catch in many developing nations is from reef fisheries. Yet, although the coral reef ecosystem is a primary biological and economic engine of Earth, these areas now face a triple threat from climate change that will probably doom many of them.[44]

Australia's 1,200-mile-long Great Barrier Reef, the greatest example of a continuous coral ecosystem and a World Heritage Site since 1981, is protected by the Great Barrier Reef Marine Park Authority (GBRMPA). About a third of the reef is now zoned as a Marine Protected Area, where fishing and anchoring are prohibited or restricted. The rest is managed in a series of special-use areas, in which activities such as fishing are regulated but permitted. Diving and snorkeling and the businesses that cater to tourists are controlled to keep damage from human recreation to a minimum. The reef, directly and indirectly, supports millions of Australians and earns more money than all of Australia's fishing industry—AU$4.6 billion (US$3.4 billion) annually for tourism-related enterprises alone.[45]

In other nations, too, the income derived from local reefs is much greater than that derived from other parts of the economy. As reef scientist Charlie Varon pointed out, the Red Sea reefs bring in more revenue for Egypt than the pyramids. Reef recreation by 18 million people in 2001 was worth $7.5 billion in Florida alone. These estimates do not include the contribution of coral to the general biodiversity of nearshore waters, such as the rich seagrass and mangrove habitats that the reefs protect. Yet coral faces many problems directly related to human activities and development, among them overfishing (the number one problem), nitrification from fertilizers, pollution from coastal cities, and destruction by dredging. These are serious threats, and coral regions such as the Caribbean have been severely damaged by some of these depredations. The fact that reefs are mostly localized may be a benefit in tackling these problems, with individual governments or regional coalitions taking the lead.[46]

Unfortunately, all the business and government capital spent on locally protecting reefs cannot protect them from the threat of warmer ocean temperatures and sea level rise. In a watershed paper

These coral reefs in the Indian Ocean, although seemingly pristine, are nevertheless vulnerable to bleaching caused by high sea temperatures. Compare these undamaged corals in the Chagos Islands to those near the Seychelles, on the following page. [FEBRUARY 2006; PHOTOGRAPH © NICK GRAHAM]

The reefs fringing the Seychelles in the Indian Ocean have failed to recover from the near-complete bleaching of 1998. Researchers reported in 2006 that the reefs, once dominated by coral, have changed to algae-covered rubble. Warming ocean temperatures guarantee continued coral damage around the world. (APRIL 2005; PHOTOGRAPH © NICK GRAHAM)

from 2003, seventeen researchers ranging from field marine biologists to biochemists to experts on ancient reefs concluded flatly that "reefs are threatened globally" and that "the link between increased greenhouse gases, climate change, and regional-scale bleaching of corals, once considered dubious by many reef researchers only 10 to 20 years ago, is now incontrovertible." Corals are limited to water temperatures between about 64°F and 89°F (18°–32°C). When temperature gets too high, even just a degree or two above summer normals, coral may expel the microscopic organisms, zooxanthellae, that, living in symbiosis with them, provide nutrients and give the coral color. When this occurs, the coral is said to have bleached. If bleaching continues for days to weeks and the symbionts do not return, the coral dies. Fleshy algae then take over the reefs, changing and limiting the ecosystem. Almost everywhere ocean temperatures are frequently nearing the bleaching point due to an overall warming of the seas. In 1998 an extreme El Niño heated tropical waters more than ever before; on the Great Barrier Reef they were the warmest ever recorded. Worldwide, coral bleaching caused death or damage to 16 percent of reefs overall and up to 46 percent in parts of the Indian Ocean. Massive corals, after surviving for seven hundred years, became lifeless rocks on the sandy sea bottom.[47]

The same El Niño also caused the die-off in 1998 of an entire coral population in Belize, which, according to scientist Richard Aronson, was "unprecedented in the past 3,000 years," based on coring studies of the ancient corals. Now in its place, he said, you see just rubble covered by algae. Caribbean reefs were bathed again in hot water during the record sea surface temperatures that hit crisis levels off Florida in August 2005 and spread south through November. Sustained heating was the greatest in twenty-one years of satellite sea surface monitoring. The water was over 86°F (30°C) for three months, and bleaching was seen at St. Croix down to 150 feet (45 m). The severest effects appeared to be in Puerto Rico and the Virgin Islands, where about one-third of the coral monitored by the National Park Service died. Elkhorn coral (Acropora palmata), recently listed as threatened under the Endangered Species Act, withstood the heat for some time but eventually half of it bleached and 15 percent died. Staghorn coral (A. cervicornis), another new addition to the threatened list, also bleached. "This was an extremely severe event," Mark Eakin, coordinator of NOAA's Coral Reef Watch program, observed; "it was unlike anything we've ever seen." NOAA reported bleaching also in the Florida Keys, southeastern Florida, and Texas's Flower Garden Banks National Marine Sanctuary in the Gulf of Mexico, as well as in Caribbean nations including the Bahamas, Barbados, Belize, the British Virgin Islands, Colombia, Costa Rica, Cuba, Jamaica, Mexico, Panama, and Trinidad and Tobago.[48]

The Great Barrier Reef and other coral reefs in the South Pacific also suffer recurring damage. Bleaching in 2002 was even worse than during 1998, affecting 60 to 95 percent of the reef, according to the GBRMPA. This time most of the affected corals recovered, but renewed high sea temperatures in 2006 affected about 190 square miles (500 sq km) and made continued recovery in some areas doubtful. "Good management allows the reefs the best chance of bouncing back," said Ove Hoegh-Guldberg, director of the University of Queensland Centre for Marine Studies. And he added, "About a third from 1998 have bounced back." But, he said, even the best management in the world cannot protect reefs from warming sea temperatures. Nor can isolation. Despite being free from the heavier damage inflicted on continental coastal reefs, the corals fringing the Seychelles in the Indian Ocean have failed to recover

CHANGING CLIMATE, CHANGING OCEAN

SYLVIA A. EARLE AND
CRISTINA G. MITTERMEIER

*National Geographic Society and
International League of Conservation
Photographers*

The ocean acts as the blue heart of our planet. Containing 97 percent of the Earth's water, the ocean yields to the atmosphere most of the vapor that then returns via clouds and rain, sleet, and snow, driving the global water cycle. The ocean stabilizes our climate against extreme swings between hot and cold temperatures. Most of the free oxygen in the atmosphere comes from organisms in the sea; without them emitting oxygen and absorbing carbon dioxide for the last three billion years or so, our atmosphere might have remained much like that of Mars. The core of Earth's life support system, the ocean governs climate and weather and keeps the planet hospitable for life in a universe where lethal extremes are normal.

It makes sense to do everything in our power to protect the systems that sustain us in the vastness of space. But widespread complacency about the stability and resilience of the natural world, especially the ocean, perhaps driven by a lack of recognition of their fragility, fosters the belief that humans have little effect on climate or the weather. Now there is compelling evidence to the contrary. The ocean is vulnerable, and changes under way in the sea—ones strongly driven by human activities—will affect not only people worldwide, but all life on Earth.

Life has prospered during times of greater cold and greater warmth. We know from evidence trapped within polar ice, for instance, that naturally changing carbon dioxide levels in the past influenced temperature, with concomitant changes in both glacial ice cover and sea level. Yet life as we know it requires a relatively narrow range of temperature and other conditions that can be—and are being—significantly altered by human actions. Even slight changes can have enormous consequences.

Ocean temperature profoundly affects what organisms can live in a given area. Warm and cold ocean currents transport a living cargo of creatures over wide areas, held in place by temperature barriers as effective as fences on land. Vertical temperature gradients have a similar function. Regions of upwelling, where nutrient-rich, cold water rises to the surface and supports impressive biodiversity, could be suppressed by changes in ocean circulation that stem from increasing temperature. While the deep sea generally stays cool, even in the tropics, surface waters are much more variable in temperature and thus more vulnerable to change. Shifts are already apparent, with some shallow tropical species venturing and settling in more temperate areas as the surface temperature rises.

Ocean circulation patterns depend on wind, the Earth's rotation, and differences in water density due to temperature and salinity, so increased temperature is liable to alter ocean currents, with consequences for temperature regimes in the sea and on land. According to the American Geophysical Union in 2003, "The average global temperature is now warmer than it has been at any point in the last twenty centuries, and, if current trends continue, by the end of the century it will likely be hotter than at any point in the last two million years." The Intergovernmental Panel on Climate Change has noted that our current rate of increase in atmo-

from the near-complete bleaching of 1998. Researchers reported in 2006 that those reefs had changed "from a coral-dominated state to a rubble and algal-dominated state."[49]

AS OCEAN WATER WARMS, it expands. This expansion, together with extra water from melting glaciers, is a major cause of sea level rise. Corals around the world, which thrive at and near the sea surface, have been able to keep up with slightly elevated sea levels in recent centuries. Although the rate of sea level rise is accelerating, it is thought that coral should be able to keep pace—as long as it remains healthy. However, not all coral grows at the same pace, and some species could be damaged by stronger currents caused by deeper water or by storms resulting from climate change.

But now there is a new concern. The ocean is the prime absorber of carbon dioxide and stores about

spheric carbon dioxide is unprecedented in the past 20,000 years. There are real concerns that increased temperature triggered by rising carbon dioxide could melt the Greenland Ice Sheet. Besides inundating some islands and coastal areas, this would reduce salinity of the North Atlantic and likely alter ocean circulation. A shift in the flow of the Gulf Stream, for example, could cause northern Europe's presently temperate climate to cool significantly. Even a small change in ocean circulation patterns anywhere in the world might initiate a chain of far-reaching consequences.

Another potentially ominous outcome of rising temperature is the release of methane from vast icy deposits in the deep sea. Fishing nets dragged through the depths sometimes catch large chunks of icy gas hydrates that quickly fizzle at the surface into water and methane gas. Methane is a far more potent greenhouse gas than carbon dioxide, and adding large quantities to the atmosphere could accelerate warming that would in turn speed the release of more methane.

Although we tend to worry about the consequences of increased carbon dioxide in the atmosphere, excess amounts of this chemical in the sea pose a problem of acidification. Carbon dioxide not consumed in ocean food webs forms carbonic acid, which lowers the ocean's pH, making the water more acidic and thus interfering with the formation of calcium carbonate. Before the Industrial Revolution, the sea surface worldwide was rich in calcium carbonate, the compound that makes up skeletons, coral reefs, and the shells of snails, mollusks, and minute photosynthetic organisms such as foraminifera, which provide significant food and oxygen. Now, higher ocean acidity has trapped an enormous amount of this calcium carbonate in ocean sediments, making it unavailable for use by shell-forming creatures. These creatures play a key role in the carbon sequestration function of the ocean, so higher acidity will also ultimately reduce the ocean's capacity to absorb carbon dioxide and to produce oxygen.

Calcium carbonate is naturally less abundant in the waters surrounding Antarctica, where massive populations of phytoplankton and pelagic snails form the foundation of critical ocean food webs. At present rates of absorption, by the end of this century carbon dioxide levels could render the Southern Ocean too acidic for these organisms to survive.

Before we can reverse these dire trends, we must recognize that a problem exists. Despite actions to safeguard critical biodiversity on land and in some freshwater ecosystems, the ocean remains neglected, with a mere fraction of 1 percent of all ocean surface waters under some form of protection (compared to about 12 percent of the land surface). The best way to forestall the harmful consequences of global warming on the ocean's biodiversity is to halt the causes of damage, protect places that remain intact or have strategic importance for conservation, and make repairs. Ocean biodiversity is the linchpin that keeps marine ecosystem services, like carbon sequestration, functioning. Lack of awareness and the resulting complacency about the importance of marine biodiversity underlie the lack of incentive to take action while there is still time. This constitutes the single greatest threat to our species.

Life will go on in some form whether or not coral reefs or rainforests exist, or whether atmospheric oxygen is reduced by half or carbon dioxide quadrupled. But maintaining living systems that can support us presents our unique challenge and will determine the legacy we can bestow on all who follow. ✦

half of all the human-made emissions (the land biosphere absorbs another 20 percent). Without this absorption, much more CO_2 would be in the air, increasing the greenhouse effect. But as the ocean takes in more carbon dioxide, it is becoming more acidic.

This leads to the third great threat to coral, which was not even guessed at a few years ago. It turns out that even a very slight increase in acidity slows the formation of calcium carbonate shells in snails, seashells, and corals. A comprehensive report by the Royal Society of London in 2005 stated, "Calculations based on measurements of the surface oceans . . . indicate that this uptake of CO_2 has led to a reduction of the pH of surface seawater." Commonly known as a measurement of acidity, pH, like temperature, is a crucial quality for coral. Oceans are normally slightly alkaline, with a historic pH of 8.15 to 8.2 near the surface (neutral pH being 7). With the added carbon dioxide, weak carbonic acid is

National Park Service scientists survey reefs in the Virgin Islands in the Caribbean during the 2005 bleaching event that damaged large sections of reef, including those protected in parks and those inhabited by endangered coral. Caribbean water was over 86°F (30°C) for three months, and bleaching was seen at St. Croix down to 150 feet (46 m). (PHOTOGRAPH COURTESY DR. CAROLINE ROGERS, USGS)

created, causing a shift in ocean water's chemistry and lowering the pH to about 8.05. This 0.1 or so may not seem like much, but on the pH scale it represents a 30 percent change. And it means that less carbonate is available to individual corals to make their shells and that some shells may be weakened in the more acidic waters. In stark terms, a more acidic ocean means coral may not be able to grow as fast or stay healthy. "This pH," reported the Royal Society, "is probably lower than has been experienced for hundreds of millennia and, critically, this rate of change is probably one hundred times greater than at any time over this period." Marine scientist Joan Kleypas of NCAR says that these are "changes in ocean chemistry that haven't occurred in 24 million years." Moreover, there is no evidence of adaptation by reef-building corals. "We don't know what will happen to individual corals, but a reef is defined by the amount of calcium carbonate. Reefs are going to decline."[50]

This is not merely a threat to coral. Levels of pH are falling even in cold Pacific waters. All sea creatures that make shells will be profoundly affected, including mollusks and crustaceans, as well as coralline algae and, more ominously, many plankton. Among the zooplankton that will be affected by a more acid ocean are the foraminifera, whose tiny shells rival snowflakes in their variety of beautiful designs. Foraminifera have been in the ocean for hundreds of millions of years; their shells persist in deep ocean sediments, carrying a record of ancient ocean conditions. The microscopic bits of calcium carbonate have created a benchmark against which we can measure present changes in the chemistry and temperature of the ocean. Of even greater importance are the phytoplankton—tiny plants—that are the basis of the ocean food web and generate half of the oxygen on Earth, while making up about half of the planet's total biological material, or biomass. "Each day," according to botanist Michael Behrenfeld of Oregon State University, "more than a hundred million tons of carbon in the form of CO_2 are fixed into organic material" by photosynthesis in these plants. This carbon enters the food chain as larger ocean creatures feed on the plankton and each other. This process is absolutely necessary for life to continue; it is also a primary way for nature to take care of excess human carbon dioxide. But greenhouse emissions could begin to change the system. Behrenfeld and his colleagues measured ocean chlorophyll using satellite imagery and charted a steady decrease in average plankton productivity since 1999, which corresponds with increases in sea surface temperature. The scientists think that warmer water mixes less well, and this limits the nutrients available for plankton. At the same time, some phytoplankton that use calcium for shells may be inhibited by the acidifying sea or even begin to dissolve in it.[51]

Scientists speculate that some bleached corals may be able to form new relationships with more heat-tolerant symbionts, but it is not known how many species can do this, how much of an impact it will have on the hardest-hit reefs, or whether the new symbionts will remain viable in the higher water temperatures expected this century. In reports in 1999 and 2004, Hoegh-Guldberg said high water temperatures and bleaching will become yearly events within three to five decades. "I think it's confirmed," he told me from his office at the University of Queensland. "The situation has probably worsened from my 1999 paper. Now we're looking at a faster pace. It appears the next few decades will see massive changes. The bleaching events that occur now in normal warm summers—relatively minor events like showing stress and losing some algae—will occur every summer to a greater degree. At the end of only twenty years we should see significant events, 50 percent bleaching, and mortality tripling." He compared the Great Barrier Reef with the Caribbean: "Now, coral does not dominate in the Caribbean; algae does. Inshore on the Great Barrier Reef it looks more and more like the Caribbean. They get hotter every summer." He calls the Caribbean "sort of a time machine, a preview of what is coming in the next years across the whole [Great Barrier] reef." The Caribbean still has many healthy reefs, but they are under grave threat. Along with disease and damage from development and recreation, increased bleaching appears inevitable. The IPCC expects that "widespread mortality" will result. Looking at the big picture, Hoegh-Guldberg said, "We are damaging a large part of the world's biodiversity" on the reefs. "We're 'chopping them down' with global warming. These reefs will be so changed that we'll have to find ways to reemploy all those people" who depend directly on reef fisheries and recreation and tourism. "The implications are huge."[52]

In 2004 Ove Hoegh-Guldberg teamed with his economist father, Hans, to suggest that Australia could save the Great Barrier Reef by changing its energy use. The two laid out a plan for reducing carbon emissions, with measures that range from renewable energy generation to a tax or trading scheme on carbon to an end to large-scale land clearing. Australia emits more CO_2 per capita than the United States and is one of the biggest land clearers in the world, 564,000 hectares (1.4 million acres, or 3,000 square miles) a year—75 percent of that in Queensland. More than 80 percent of Australia's power comes from coal, and its power plants alone emit more greenhouse gases than nations such as Argentina and Sweden contribute from all sources. Ian Campbell, Australia's minister of environment and heritage at the time, acknowledged climate change, but his Coalition government did not ratify the Kyoto Protocol. National elections brought Labor to power in 2007, and Australia quickly joined Kyoto.[53]

The Great Barrier Reef, near Townsville, Australia, part of the irreplaceable coral habitat that provides a livelihood for 500 million people around the world. Oceans have taken up about half of the carbon dioxide produced by humans, which is making them more acidic than they have been in 24 million years. This increased acidity will impede corals and other ocean creatures from making their shells. (FEBRUARY 2005)

Australia has more than its World Heritage coral reefs to worry about. A severe drought has gripped northeastern regions since 2002, and although record-setting storms can dump unprecedented rainfall, they do little to replenish soil moisture. Queensland's annual rainfall is now 5 to 10 percent lower than it was thirty years ago, yet water use has doubled in the past fifteen years. The year 2005 was the hottest ever recorded in Australia by the Bureau of Meteorology. While driving in Queensland in February 2005, I saw a brown line along the western horizon quickly thicken into a wall of reddish brown dust, roiling like a thunderstorm. Just after I pulled off to photograph it, it enveloped me in a gusty, gritty haze for more than an hour. The giant dust storm of February 2 and 3, 2005, was up to 600 miles (1000 km) across and driven by gusts of up to 50 miles (80 km) an hour. It blew an estimated 4.8 million tons of dirt across Queensland, closing twenty airports. At the same time, intense storms hit up and down the east coast of Australia, dropping record rains, including the highest twenty-four-hour rainfall ever recorded in Victoria. In 2008 the Bureau reported, "The combination of record heat and widespread drought during the past five to ten years over large parts of southern and eastern Australia is without historical precedent and is, at least partly, a result of climate change." Bureau of Meteorology scientist Mark Williams was quoted as saying the powerful storm was a "once-in-a-lifetime phenomenon." Wildfires in eastern forests flamed so intensely, wildlife officials reported, that endemic birds and frogs probably were pushed to extinction.[54]

CHINA, THE EMERGING GIANT

China is the new world heavyweight, not only in the global economy, but also in climate change; with the United States, it looms across borders to affect the entire planet. As a developing nation, it stands outside the guidelines of the Kyoto Protocol, yet with more than 1.3 billion people and a huge energy-gobbling economy, it is first in coal consumption and in total carbon dioxide emissions, having just exceeded the United States. China recently passed Japan to become the world's second-largest consumer of oil, burning about 7.2 million barrels of oil every day, twice what it produces. Many of its cities suffer from choking air pollution, and large regions are beset with drought, failing crops, and sandstorms linked to global warming. Even though China's leaders remain fixed on rapid development and increasing energy use, the first steps are being taken toward emissions control and alternative energy sources. It is mandatory to try to understand China if one wants to have a world view of global warming—and it is crucial to include it in plans for reducing the effects of climate change.[55]

The instant I boarded a bus from the airport to Beijing's city center in early 2005, the disparities of wealth and living conditions in China started coming clear. The multimillion-dollar skyscrapers and jammed freeways give central Beijing the feel of Los Angeles or Houston. For miles around, however, low-slung houses, monotonous apartment blocks, and pedestrian-filled streets provide stark contrast. There are hundreds of working-class neighborhoods, like one I visited near a huge steel plant on the west side of town. I was struck by the relative lack of bicycles in the city center, defying the stereotype. Shanghai and Guangzhou (formerly Canton) also are metropolitan centers with millions of citizens, freeways, and glittering central districts. The development of these huge cities and of a growing urban middle class has been fueled by a voracious appetite for energy—more than two-thirds from coal—with little attention to pollution control. The result has been frequent smog in the cities and widespread environmental degradation across the country. In recent years, China has also begun to feel the sting of climate change.

I flew to Guangzhou, then motored north on modern highways to the smaller city of Shaoguan in rural Guangdong Province. Here, an older image of China sprang to life in streets crowded with bikes and trucks. I had come to the Shaoguan region with campaigners from Greenpeace/China to witness a severe drought that had gripped the area for six years. Forty million people were affected, and nine million had difficulty getting drinking water. Economic losses due to rice and other food crop failures amount to more than 6 billion RMB ($US748 million). To the north of Guangdong, a band of 68,000 square miles (176,000 sq km) extending through seven provinces all the way to Inner Mongolia was also parched. This drought in 2006 covered a 21 percent larger area than the one the year before.

With the modern prosperity of Beijing and Guangzhou fresh in my mind, I was struck by a rural life that by comparison is ancient and poor. Forty-seven percent of Chinese live on less than two dollars a day. Hundreds of millions of people still live in villages or towns built of native stone or brick, surrounded by all-important rice paddies and fields that take up every inch of flat land. It was wrenching to see this already tenuous way of life drawn out thin and ragged by lack of water. I met beekeepers whose insects were starving because there were no flowers, villagers who had to carry water in shoulder buckets more than half a mile (1 km), farmers who had to buy rice because their crops had failed; once-lush forest plantations that were now scorched by fire and drought.

In Daqiao village, a fifty-five-year-old man said the river that flowed under the picturesque arched bridge for which the village was named was lower than it had ever been during his lifetime. The rice crop had failed, he said, and a replacement planting of cabbages was also drying up. Taps in the town water system had gone dry, so residents were dipping water out of the trickling river, making their way past festering piles of garbage that tumbled down the banks. Here the drought was revealing the short-sightedness of the village's easy but unsanitary waste disposal system, which relies on the river rising up and taking the town's garbage each spring; yet now the rains no longer come to fill the river.

Everywhere rice paddies turned to dust while the grain was still developing, and after that fish ponds and wells that had been relied on for hundreds of years went dry. Guo Konzeng, a sixty-six-year-old in a village named for his family, said it was the driest year he had ever lived through as he pointed out his six empty granaries. He had harvested not a single grain of rice. Guo's village also was threatened by a forest fire that destroyed a pine plantation on the surrounding hills. He showed us where the flames were extinguished just in time at the edge of his settlement.

In the Qingyuan area at the northern edge of the Pearl River delta, rainfall in 2004 was the lowest on record, and the drought was drawing down reservoirs and reducing even major rivers to meandering sloughs among the sandbars. The Beijiang (North) River near Qingyuan City reached an all-time low in winter 2004–5. Usually a major transport artery, the shrinking river trapped more than one hundred ships and barges until water was released from a dam. Grandiose floating seafood restaurants along the banks sat tilted on the sand.[56]

This drought is not a fleeting, unusual event in Guangdong. Whereas between 1950 and 1980 the region experienced only four droughts affecting more than 2 million acres (800,000 Ha) of land, in the

A woman in northern Guangdong Province, China, carries
water past an evaporated
fishpond, her well having gone
dry during a severe drought.
(JANUARY 2005)

In Daqiao, northern Guangdong Province, a woman washes clothes in a polluted river, which has dwindled to a very low level during a drought in 2005.

past quarter century such droughts have occurred eleven times. Jia Tianqing, deputy director of the Guangdong Meteorological Bureau, told Greenpeace/China that the province is undergoing a remarkable warming, as indicated by one hundred years of records.

The same is true of China as a whole. Water shortages are growing throughout the country; according to one estimate, some 60 million people are having difficulty finding the water they need. The Beijing area, home to more than 17 million, is suffering the worst drought in fifty years and continues to choke under thick, gritty dust blowing in from the west. In spring 2006 the worst of a series of ten sandstorms turned skies yellowish orange with 300,000 tons (271,000 mt) of sand coming from the direction of Mongolia. The dust was easily visible on satellite images—brown smudges moving east to Korea, Japan, and on across to North America. Drought has become endemic in northwestern China, in Hebei and Inner Mongolia, among other provinces. Late in 2006 key hydropower reservoirs had 12 percent less water than in 2005, forcing the use of more coal-powered plants to generate electricity. Even the Huang He—the fabled and crucial "Mother River," the Yellow River—known in the past for huge floods, has shrunk in its bed to an intermittent flow, strangled from its very headwaters in Qinghai Province on the edge of the Tibetan Plateau. One visitor found that the source lakes "had shrunk to cracked dry beds stretching to the horizon." A Greenpeace/China expedition to the region in 2005 found glaciers retreating ten times faster than in the twentieth century, permafrost thawing, and precipitation declining while temperatures soared—a trend that has been on the rise since 1986.[57]

The effects of China's industrial growth, population, and fossil fuel use are not restricted to the countryside. Back in Beijing, out past the west end of Subway Line 1, I walked between steaming, thundering steel mills and was suddenly confronted with a paradox of time and culture: an ancient pagoda. Hundreds of years ago, the rivers west of the Forbidden City and southwest of the Summer Palace ran through low hills. During the Ming Dynasty, Shijing Pagoda was built on 600-foot (184 m) First Fairy Mountain, overlooking this rural landscape. Since the early twentieth century, though, this area has been an industrial suburb, overshadowed by the cooling towers, furnaces, and smokestacks of the mill compound, which covers many square kilometers. The Shougang Group, known as Capital Steel, eventually grew right around the pagoda, largely obscuring it from view.

Capital Steel employs about 240,000 workers and is Beijing's largest industrial polluter, annually discharging about 18,000 tons (16,200 mt) of particulate matter—40 percent of the entire urban industrial-sector emissions. Recently, under pressure economically and pushed by the government ahead of the impending 2008 Olympics, the company has sharply reduced its emissions and begun moving operations to Hebei Province. Of course, this shift of the pollution burden to another city is a transparent gambit. Sixteen of the twenty most polluted cities in the world are in China, according to the World Bank, including Beijing, Guangzhou, Shanghai, Xi'an, and Shenyang. The World Bank calculated in 2003 that many cities had unhealthy levels of sulfur dioxide (SO_2) or nitrogen dioxide (NO_2), which cause acid rain. China, in fact, sends up more SO_2 than any other nation, and in 2005 emissions were more than 27 percent over 2000 levels, according to government figures. It is widely estimated that more than 750,000 Chinese die of air pollution–related diseases each year. China is the world's largest mercury polluter as well.[58]

Capital Steel, the largest steel manufacturer in Beijing and the fourth largest in China. This plant, a severe polluter and carbon emitter on the city's west side, cut production in preparation for the 2008 Olympics. (JANUARY 2005)

Kids play on a sand pile in a back street of a town dominated by the 3,000 megawatt Dalate coal-fired power plant, just south of the Huang He River, near Baotou, Inner Mongolia. Hundreds of new coal-fired power plants are being built in China, threatening to overwhelm efforts to control carbon dioxide and unhealthy pollutant emissions. (JANUARY 2005)

This land has undergone a withering attack from hundreds of years of development that leveled forests, carved out coal mines, constructed dams, and pushed farms and cities across river bottoms and estuaries. During Mao's Great Leap Forward in the 1950s, an estimated 59,000 coal mines were opened, feeding 600,000 steel furnaces. Pollution was uncontrolled. Forests were leveled and replaced by farms that were poorly managed. Famines resulted that killed more than thirty-five million people. Conditions did not improve during the Cultural Revolution, and serious environmental disasters have continued to occur in more recent years. Forests have been reduced to less than 17 percent of total land area from once having covered more than half. Ninety percent of grasslands are degraded from overgrazing and climate change, and desertification is rampant. The state environmental agency estimates that more than a thousand plant species and nearly five hundred birds and other animals are under threat of extinction.[59]

Low-grade, high-polluting coal generates more than 75 percent of China's electricity, according to the Natural Resources Defense Council (NRDC), and over the next twenty-five years nearly as many coal-fired plants will be built in China as in the rest of the world combined. Leaders are determined to quadruple the economy in just fifteen years. Development is so rapid that China overtook the United States as the world's leading carbon dioxide gas emitter in 2008. Up to 20 percent of China's power plants may be unlicensed and polluting at will. The more than two billion tons of coal mined and burned is responsible not only for carbon emissions but also for the more visible smog, particulates like black carbon soot, and nearly all of China's sulfur dioxide emissions and resulting acid rain. In addition to being consumed in huge quantities by steel mills and power plants, coal is used by individuals throughout the nation for heating and cooking; in farm yards, a regular chore is pounding wet coal dust into molds to make coal cakes called "beehive coal."[60]

The coal mines themselves are enormous sources of CO_2 because of more than fifty unquenchable fires burning in them. They exact a huge human toll, too: more than six thousand miners were killed in Chinese mines in 2004 and 2005, and the high death rate continued into 2006 (unofficial death tolls are 40 percent higher). Many of these deaths are in smaller mines, which the government has said it has begun to close down.[61]

If coal is the historic source of China's pollution, the coming menace is from vehicle emissions. Although the number of vehicles of all types per person is still very low, the overall number has surpassed 26 million and is growing by more than 10 percent a year as the expanding middle class invests in private transportation. China predicts 140 million private cars on the road by 2020, about half the U.S. fleet—though Chinese vehicles will get better mileage, at least under current laws. About 356,000 more cars merge into Beijing traffic yearly, for a total of about 3 million—one car for every five residents. Shanghai's 2 million cars create 70 to 80 percent of that city's air pollution—this despite 500 miles of subway. China's demand for petroleum (which also fuels ever more power generators) is growing as a result at more than 10 percent a year; in 2004 alone, the growth rate was 15 percent. China is now the second largest oil importer and increasingly vies with the United States for the output of the world's oil fields.[62]

The damage that China inflicts on the Earth's atmosphere is already immense, and with more than forty to fifty coal power plants being built each year, it is going to get much worse. (The Chinese government is also planning to dam more large rivers; while cleaner than coal, hydroelectric power exacts enormous environmental costs: the Three Gorges Dam already has shut off so much fresh water to the Yangtze Delta 2,000 miles [3200 km] downstream that plankton needed by the nation's rich coastal fishery are in rapid decline, and sediment to keep the delta healthy is down by 55 percent.) Still, China could change direction and turn its growing wealth and industrial might toward cleaner technology, renewable resources, and a vision that looks beyond the template set by the West. Though not required to do so, because of its status as a developing nation, and despite having previously resisted international pleas to set emissions reduction goals, in 2002 China ratified the Kyoto Protocol. It reportedly had reduced its carbon emissions total in the late 1990s. However, its power generation increased by more than 12 percent in 2004, and fossil fuel use in 2006 was 9 percent above 2005. Although coal use is increasing faster than anywhere else in the world, some positive steps are being taken. The State Environmental Protection Administration halted construction of twenty-three new power plants, citing lack of approval of their environmental impact. Some two hundred factories in Beijing are being closed, and the city has strict controls on industrial smoke, automobile exhaust fumes, and construction dust. New national taxes aim to curb the use of larger gas engines and certain wood products, including chopsticks. By 2020, China intends to double the amount of electricity produced by renewable sources. It is already the second-ranked solar cell maker and is nearing the lead in export of wind turbines. China will try to attract large amounts of Clean Development Mechanism money through the Kyoto Protocol. So far, however, it has failed to meet initial efficiency targets.[63]

Qin Dahe, chief of the China Meteorological Administration and coauthor of an initial report on the IPCC 2007 assessment, said that China takes global warming "extremely seriously." But at the same time, warned a foreign ministry spokesperson, developed nations have long added to greenhouse emissions, and their responsibilities "cannot be shirked." Plus, one-third of China's emissions come from manufacture for export, mostly to those very nations and for products marketed by leading corporations. As China's pollution threatens to negate any emission controls under the UN Climate Convention, there are strong reasons for the West and Japan to collaborate with her on carbon reductions and alternative energy schemes.[64]

If global warming, together with the serious changes it brings to air, water, plants, animals, and people, is to be brought under control, China must choose a path that diverges from its recent history of relentless, heedless development. (This is true for India, too, which is poised to surpass China in population very soon.) It is a path poorly marked by China's Western competitors and trade partners. Those nations—the United States especially—must likewise turn to reinventing and reimagining the future, not just on a national level, but for the world as a whole. If we don't, we will surely fail, and the world that belongs to us all, that gives us all that we need and value, will begin just as surely to fail us.

CHOOSING A SAFER, CLEANER, AND COOLER WORLD

t Is Five Minutes to Midnight." With that headline in January 2007, the *Bulletin of the Atomic Scientists* moved its clock two minutes closer to doomsday. And for the first time the scientists concluded that "the dangers posed by climate change are nearly as dire as those posed by nuclear weapons."

In more than a hundred years of making choices about our economy and how to power it we have sought abundant energy and improved the lives of many millions. Now we are learning that those energy sources, primarily coal and oil, generate most of the excess heat that is flowing into the atmosphere. Global warming has been called a time bomb because of the way heat energy builds up and sets into motion changes that take many years to play out. It has also been called a hoax. For a long time, for many reasons, most of us didn't see it coming. As Dean James Gustave Speth of the Yale School of Forestry and Environment wrote in 2004, "We are tragically late in addressing climate change; irreparable damage will unfold in the decades ahead due to our past negligence."[1]

CHILLING FORECAST FOR A WARMER WORLD

When the Group of Eight most developed nations (G-8) met in 2008, they agreed for the first time to voluntarily make "deep cuts" in greenhouse gases. Previously, President George W. Bush had refused any reductions. But at the 2005 G-8 meeting leaders got a succinct but chilling forecast. "The cost of failing to mobilize in the face of this threat is likely to be extremely high," stated a report called "Meeting the Climate Challenge." "The economic costs alone will be very large. . . . The social and human costs are likely to be even greater, encompassing mass loss of life, the spread or exacerbation of diseases, dislocation of populations, geopolitical instability, and a pronounced decrease in the quality of life. Impacts on ecosystems and biodiversity are also likely to be devastating."[2]

We've been feeling the effects of Earth's changing climate since the world began to industrialize and intensify its use of fossil fuels. Since the late nineteenth century, the atmospheric concentration of carbon dioxide has increased by 36 percent, to its highest level in more than 800,000 years, and levels are building two hundred times faster than at any previous point in history. The level of methane is also a great deal higher, as is that of several other greenhouse gases. In recent years, average air temperature has risen more than 1.4°F (0.8°C); surface waters of the oceans have warmed by a degree; and the rate of average sea level rise has nearly doubled. The rate of temperature increase has more than doubled since 1975. There is almost no possibility that all this has happened without the involvement of human-generated greenhouse gases.[3]

Just how much more the temperature will go up is the subject of intense scientific study—and political interest. The basic facts of atmospheric warming are known, and they have to do with the amount of greenhouse gases and how they interact with the rest of the atmosphere. But the complexities of the air and its effect on land and sea are staggering. Some variables can't be easily pinned down, starting with the effects of clouds and dust on atmospheric warming and ending with human factors like evolving technology and emission standards. These complexities have challenged the hundreds of scientists

PREVIOUS PAGE Wind turbines in Ijesselmeer, the Netherlands. Wind alone, if properly harvested, could easily supply all the world's electrical power. Generating capacity is expanding by more than 25 percent each year. (AUGUST 2004)

charged with making future estimates for the 2007 Intergovernmental Panel on Climate Change Fourth Assessment (IPCC AR4). They've managed the variables using twenty-three computer programs that have proven skillful at modeling the workings of the atmosphere, including being able to "hindcast"— perform a test in which real data from the past century, like the amount of greenhouse gas, is entered into the program to see if its results match the climate that actually occurred. Climate models can do this with great accuracy now, right down to seeing the dip in temperature from volcanic eruptions like that at Mount Pinatubo. To calculate potential future climate, forty scenarios of possible human development in this century are used (originally devised for the 2001 report), involving CO_2 levels lower than the current point up to levels that are seven times the current amount. Then, with changing scenario variables, the climate modeling programs estimate future temperature, rainfall, and other climate attributes. Detailed testing and verification of the computer programs, along with thousands of new field observations, are significant advances over the 2001 IPCC. Through this painstaking analysis, the scientists have increased the evidence of climate change while reducing uncertainties about its effects and meaning.[4]

These new findings support what scores of research papers published in the past few years have been telling us—that Earth's temperature is going to continue to rise steeply through this century unless the amount of greenhouse gases sent into the air can be greatly reduced. If the current rate of warming continues, by 2100 the world will be about 3.2°F (1.8°C) hotter. If human society follows a path this century toward headlong development with little control of emissions, the Earth could roast, reaching a temperature more than 7°F (4°C) hotter than today. That is a lot of heat for the atmosphere to gain in a mere hundred years, considering that right now the Earth is only ten to twelve degrees warmer than it was at the depth of the last ice age, 20,000 years ago.

How quickly greenhouse gases are put into the air also matters. The point at which they reach concentrations twice as high as those before industrialization is an important benchmark for scientists. That level is about 560 ppm CO_2 (a figure that includes all greenhouses gases and other atmospheric influences on temperature, though for simplicity it is expressed in terms of CO_2), and it is one we might see as early as 2050 if emission levels don't change. This is likely to result in an average world temperature almost 4°F (2.2°C) warmer than today's. Some scientists are warning, however, that their calculations already forecast more emissions from warming soil and less uptake by the ocean, factors that would push temperatures higher sooner.[5]

HITTING THE BRAKES

The purpose of these warming projections and the IPCC reports is to help scientists, politicians, and all of us establish a danger point that we will try to stay well below. The Earth's atmospheric momentum is very large, and some future heating is already inevitable. How much more can we risk?

The central question is, since the changes described in this book are the result of less—sometimes much less—than 1.4°F (0.8°C) of heating, what will happen as we continue to raise the Earth's temperature by the three and a half degrees the current rate of warming portends? According to the IPCC 2007 assessment, even if we had shut off all greenhouse emissions in 2000, what we have already put into the system will raise temperatures by 1.5°F (.9°C) by 2100. Thus greater environmental changes are certain to happen within the lifetime of a child born today. The UN Framework Convention on Climate Change, the international climate treaty of 1992, is intended to prevent change that could dangerously disrupt ecosystems and sustainable human endeavors. Yet the tipping point remains undefined. Some climatologists believe we've already hit it. And most guess that dangerous disruption will occur when the average temperature climbs 1° to 3°F (0.6°–1.8°C) higher than it is now. Increasingly, scientists think the danger is here and our course must be reversed.[6]

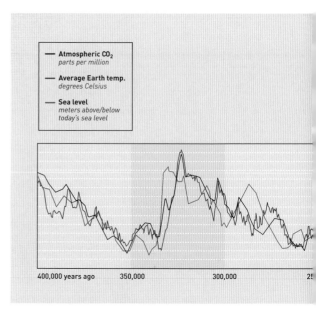

If trends continue unchecked, some of the changes already documented will intensify in the coming century and will surely be dangerous. Higher tides and stronger ocean storms devastating more major cities. Total loss of coral reefs in the Caribbean or Coral Sea. Complete melt of Arctic summer sea ice and deep thawing of permafrost. The slowdown of the great currents of the Atlantic. Giant floods and outbreaks of disease. Overwhelming drought and heat in Africa and China. The trigger for these could be a simple event, such as a sudden release of methane from the Arctic or the seafloor, that crosses a threshold we may not realize we are approaching. The result would be disruptions in food supplies, transportation, employment, and the ability of governments to respond to people's needs. The number of environmental refugees fleeing coastal areas and weather disasters—and not just in poorer nations— would double, according to one prediction, to 50 million people by 2010. Humanitarian goals agreed to by the United Nations are already at risk of failure. These are effects that will change our rather benign-sounding terminology to "global *climate disruption*" and "climate *meltdown.*"[7]

It is hard to imagine anyone not wanting to head off these changes. One of the grimmest warnings came before the G-8 in 2005. The report to them put the "dangerous" level of future temperature rise at about 2°F (1.2°C). It warned that unfettered industrialized development, because of what was being put in the air, would undermine the world and its poor. Low- and zero-carbon technologies were needed immediately. That same year, 1,300 scientists and experts participating in the UN Environment Programme's Millennium Ecosystem Assessment said that many parts of the natural systems that support life on Earth are worn and depleted. We've overused them. Nearly two-thirds of ecosystem services such as nutrient recycling and air cleansing, as well as fresh water and food, supplied at no cost to us, were found to be in decline. The report focused in particular on the effects of global warming and changing atmospheric composition. "The change with the greatest potential to alter the natural infrastructure of Earth," it said, "is the chemical experiment humans have been conducting on the atmosphere for the past century and a half." Scientist James Lovelock, who theorized Gaia, the living, self-regulating Earth,

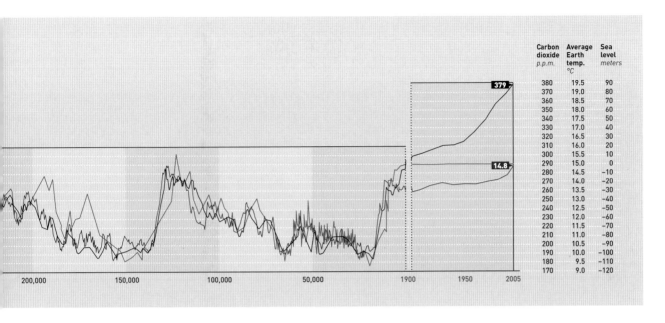

Carbon dioxide p.p.m.	Average Earth temp. °C	Sea level meters
380	19.5	90
370	19.0	80
360	18.5	70
350	18.0	60
340	17.5	50
330	17.0	40
320	16.5	30
310	16.0	20
300	15.5	10
290	15.0	0
280	14.5	−10
270	14.0	−20
260	13.5	−30
250	13.0	−40
240	12.5	−50
230	12.0	−60
220	11.5	−70
210	11.0	−80
200	10.5	−90
190	10.0	−100
180	9.5	−110
170	9.0	−120

Atmospheric CO_2 concentrations, estimated temperatures (from Antarctic ice cores), and estimated sea level over the last 400,000 years, with the last 105 years of average readings shown in detail. Previous temperature cycles were influenced by subtle orbital changes over thousands of years, but atmospheric CO_2 played a major role. Scientists are concerned about the high level of CO_2 today because of its unquestioned heat-trapping ability. (ADAPTED FROM DAVID TALBOT, *TECHNOLOGY REVIEW* [2006]:40–41)

says her healing will take many thousands of years, as the support systems are failing under a "morbid fever."[8]

Even though a few scientists are deeply pessimistic about the chances of reversing this situation, many of the more experienced continue to be confident that cutting greenhouse emissions will make a great difference. One of the strongest scientific voices of warning is James Hansen, lead climate scientist at NASA's Goddard Institute for Space Science and one of the first to bring long-term climate change information to Congress and the public. "The dominant issue in global warming, in my opinion," he wrote in *Scientific American* in March 2004, "is sea-level change and the question of how fast ice sheets can disintegrate. A large portion of the world's people live within a few meters of sea level, with trillions of dollars of infrastructure. And once large-scale ice-sheet breakup is under way, it will be impractical to stop. Dikes may protect limited regions such as Manhattan and the Netherlands, but most of the global coastlines will be inundated." In 2005 he cautioned that "a 20-m sea level rise is not required to wreak havoc with civilization today. Three-quarters of a meter each from Greenland and Antarctica would do the job quite well."

Hansen and his co-researchers also see that today's greenhouse gas concentrations are "far outside the ranges that existed for hundreds of thousand of years." Earth is absorbing nearly one watt of heat energy per square meter more than it radiates back to space, and much of it goes into the sea. He thinks our target for CO_2 levels should be 350 ppm—not seen since the 1980s. Yet Hansen advocates strongly

The Thames barrier against high tides, built in 1982 and now used six or seven times a year to protect London from flooding. Coastal cities around the world will have to build barriers such as this one and the dikes of Holland to adapt to sea-level rise and stronger storms. (JULY 2004)

for an alterative scenario, "practical actions with multiple benefits," which can keep global warming below 1°C. He advocates, along with CO_2 cuts, seeking early reductions in methane, nitrous oxide, and even black soot. He warns that we are on a "slippery slope"—the longer we wait to dramatically slow the greenhouse effect, the greater the chance that the heat built up in the air and ocean will bring about dangerous changes.[9]

The Klaxon is sounding. Dire and detailed forecasts of a world 3°F or more warmer than today, and the benefits of not only avoiding the most dangerous changes but also creating a cleaner world, have brought many ideas and technologies to the forefront. We will explore some of these later in this chapter. Meanwhile, it is important to realize that there may be surprises, and to keep in mind how Wallace Broecker of Columbia University's Lamont-Doherty Earth Observatory summed up both our hubris and our dilemma: "The climate system is an angry beast and we are poking at it with sticks."[10]

OUR PLANET ON CO₂

Individually, we do not feel very much in control of the world— certainly not in a position to force the climate to do anything, with or without a stick. But the extra heat the planet absorbs can be traced directly to the excess greenhouse gases humans are putting into the air. There is no doubt about this because the carbon released from fossil fuels can be detected and measured. Perhaps rather than a sharp stick, we should think of the tool in our hands as a lever and the atmospheric balance of heat and radiation as a huge boulder that we have pried up just enough to start it rocking, without knowing exactly where the toppling point is. In the atmosphere, we wield our influence—our lever—primarily through the burning of fossil fuels for power, heating, cooking, chemical making, and transportation, thereby raising temperatures.

Each year the world vents into the atmosphere more than 32 billion tons (29 billion mt) of CO_2 from oil, gas, and coal. Carbon dioxide from the manufacture of cement, together with other gases like methane, nitrous oxide, and ozone-depleting gas substitutes, contribute the equivalent of more than 14 billion tons (13 billion mt) of CO_2. And billions of tons more carbon dioxide and methane are released in cattle and sheep ranching, when forests are cut, when fields are burned and plowed, when rivers are dammed, and when urban areas are paved and roofed over. If all the carbon in this yearly emission of greenhouse gas (14 billion tons) were anthracite coal, it

Coal is the most abundant fossil fuel on Earth, and burning it generates almost 40 percent of the greenhouse gases emitted by the energy sector. Per unit of energy, coal releases a third more carbon than does oil. Success in limiting greenhouse gases depends on using less coal, burning it much more efficiently, and keeping the CO_2 that it produces out of the atmosphere. (CHINA, JANUARY 2005)

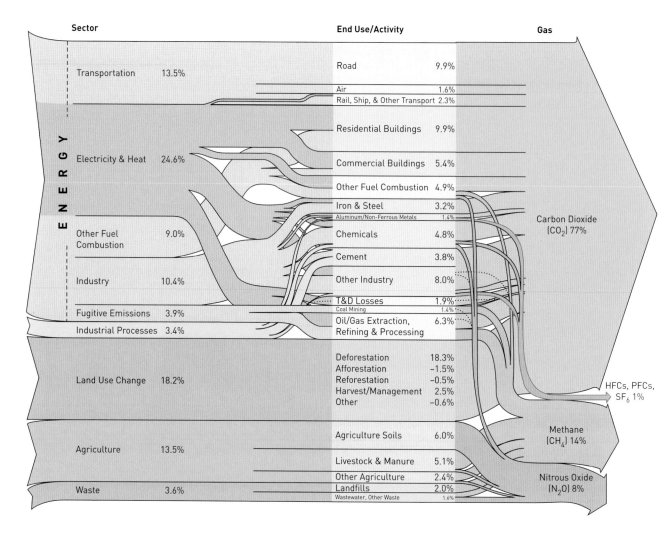

Sector		End Use/Activity		Gas

Sector

Transportation   13.5%

E N E R G Y

Electricity & Heat   24.6%

Other Fuel Combustion   9.0%

Industry   10.4%

Fugitive Emissions   3.9%

Industrial Processes   3.4%

Land Use Change   18.2%

Agriculture   13.5%

Waste   3.6%

End Use/Activity

Road   9.9%
Air   1.6%
Rail, Ship, & Other Transport   2.3%
Residential Buildings   9.9%
Commercial Buildings   5.4%
Other Fuel Combustion   4.9%
Iron & Steel   3.2%
Aluminum/Non-Ferrous Metals   1.4%
Chemicals   4.8%
Cement   3.8%
Other Industry   8.0%
T&D Losses   1.9%
Coal Mining   1.4%
Oil/Gas Extraction, Refining & Processing   6.3%
Deforestation   18.3%
Afforestation   -1.5%
Reforestation   -0.5%
Harvest/Management   2.5%
Other   -0.6%
Agriculture Soils   6.0%
Livestock & Manure   5.1%
Other Agriculture   2.4%
Landfills   2.0%
Wastewater, Other Waste   1.6%

Gas

Carbon Dioxide (CO_2) 77%

HFCs, PFCs, SF_6 1%

Methane (CH_4) 14%

Nitrous Oxide (N_2O) 8%

Sources of total gas emissions for 2000. Of the greenhouse gases released, 35.2 billion tons were CO_2 and 10.7 billion tons were other gases. By 2004, annual emissions had risen to about 42 billion tons of CO_2 and 14 billion tons of other gases. Almost every modern human activity, from car travel to waste disposal to farming, results in the emission of gases that trap heat in the atmosphere. (ADAPTED FROM *NAVIGATING THE NUMBERS*, WORLD RESOURCES INSTITUTE, 2005)

would make a pile about three miles in diameter and a mile high. The United States, with less than 5 percent of the world's population, contributes almost 16 percent of the total heat-trapping emissions each year (and 23 percent of the CO_2 generated by the burning of fossil fuels alone). China came in second as of 2004 official UN figures, but is thought to have passed the United States by 2008. Indonesia and Brazil together contribute 12 percent, thanks mostly to their land-clearing and burning practices. Russia emits 4.8 percent of global greenhouse gases, India 4.5 percent, and Japan 3.2 percent. And finally, the twenty-five (in 2006) European Union nations accounted as a unit for about 11.5 percent of

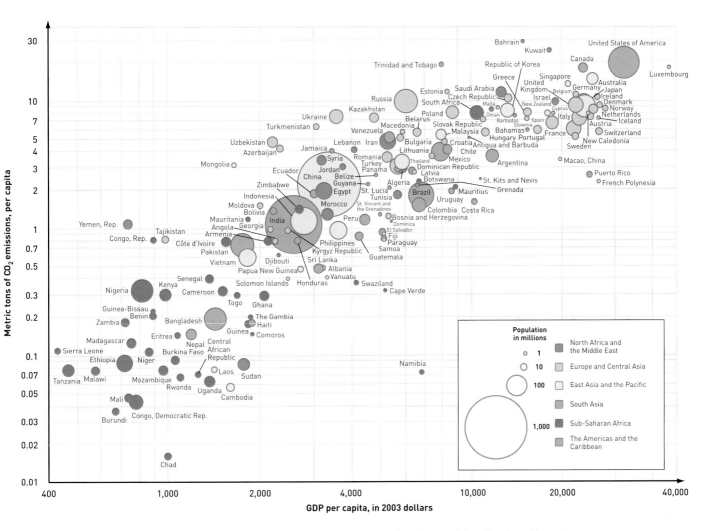

A representation of 182 countries, showing relative wealth and CO_2 emissions, based on 1999 data. The disparities are larger than they appear on the logarithmic graph: the U.S. has well over thirty times the gross domestic product per person of Chad, and a thousand times more per capita emissions. [ADAPTED FROM HANS ROSLING, GAPMINDER FOUNDATION, WWW.GAPMINDER.ORG. DATA FOR YEAR 1999 FROM WORLD DEVELOPMENT INDICATORS, WORLD BANK, 2003]

total emissions. The rest—about 35 percent of the world total—is concentrated in the Western and developed nations, including South Korea, Canada, and Australia. Most of the 191 UN member nations have very low emissions.[11]

The American transportation sector, which depends almost entirely on oil, chugs out about a third of the national CO_2 total. There are 231 million automobiles, vans, pickups, and SUVs in the United States. If an American is lucky—or concerned—enough to own a vehicle that gets 30 miles to the gallon, driving 10,000 miles over the course of a year will cough out more than 3 tons (2.7 mt) of CO_2.

Oil refinery, Carson, California. Americans spent $1 million a minute (in 2008) on imported oil, and much of this imported oil is refined into gasoline within the United States. The U.S. transportation system, which depends almost entirely on oil, chugs out about a third of the national CO_2 total. There are 242 million automobiles, vans, pickups, and SUVs in the United States. (MARCH 2005)

Many SUVs and pickups get only half that mileage, as their owners are no doubt acutely aware. The electrical part of American energy generation accounts for about another 40 percent of total U.S. carbon emissions. All in all, the United States relies on fossil fuels to meet more than 85 percent of its energy needs: 40 percent from oil, 23 percent from natural gas, and 23 percent from coal. Eight percent of power is nuclear, and the rest is generated from various renewable sources.[12]

THE TRUE COSTS OF FOSSIL FUELS

This great dependency came about because coal and petroleum are incredible sources of energy from the ancient Earth—powerful, portable, and plentiful. Humans have used coal for thousands of years. It became a crucial energy source in the Middle Ages, and despite its acrid and polluting fumes (which caused it to be outlawed several times), it fueled the Industrial Revolution. Much more widely available than oil, coal produces 40 percent of the world's electricity and is the prime fuel for many industrial plants worldwide.

Petroleum, too, has been known since antiquity, but the process of distilling it was invented only about 150 years ago. Since World War II we have refined and reformulated petroleum to produce fuels, fertilizers, and plastics—and with them, we've created the modern world. In a recent article, *Ecologist* magazine listed more than 160 petroleum products, from adhesives and air conditioners through medical equipment, movie film, water pipes, and wax paper. Oil made possible the invention of the automobile and the airplane. Oil also changed the face of warfare and, some would say, sowed the seeds of modern terrorism.

World petroleum demand increased sharply by 3.2 percent in 2004, though in 2006 the rate of increase declined to below 2 percent. At an annual increase of 2 percent per year, by 2040 the world will have to pump more than 160 million barrels a day—twice as much as it does today. Very little of this petroleum will come from the United States, which has less than 3 percent of world reserves, yet sucks up 25 percent of world crude. China uses more than 8 percent of world production, and its demand is growing by almost 16 percent a year. Yet the point at which oil will become less plentiful is near at hand. By nearly every estimate, including those of the oil industry and petroleum-producing nations, the peak of crude production—at which point the rate of oil production on Earth will enter a terminal decline—is getting close.

To keep using oil farther into the future, the world will have to try to pump it out of every possible location; it will also need to turn to oil sands and shales, which involve extraction processes that require large amounts of energy and are devastating to landscapes. Natural gas, which is methane from petroleum wells and increasingly from separate drilling, is a mainstay for much industrial and residential heating and has been substituted for coal to generate electricity. The United States is much more independent in gas than in oil, importing only about 15 percent of its natural gas, mostly from Canada. The world's greatest reserves of natural gas are in Russia, Iran, and Qatar. In recent years the price has skyrocketed in the United States, causing gas to become less economical for energy production. At best it is a temporary substitute for most of our oil; not only will it, too, peak not long from now, according to analysts, but it is also much less useful as a transport fuel. The major benefits of natural gas with

ABOVE The Belridge oil field near Bakersfield, California, has yielded more than a billion and a half barrels of petroleum since the first well was sunk in 1911. A thousand new oil and gas sites are being drilled or planned today. (JUNE 2005)

BELOW Kayford Mountain, near Charleston, West Virginia, is being disemboweled by coal-mining activities. About 1,000 feet of the once richly forested mountain have already been blasted and scraped off into surrounding valleys. The man pictured is Larry Gibson, whose family owns land on the mountain. He opposes continued mountaintop removal mining in the Appalachian Mountains. (SEPTEMBER 2005)

respect to global warming are that it produces about 30 percent less carbon dioxide than the equivalent heat value of oil and that it can be produced from the methane in landfills and agricultural waste.

Most alternative sources of energy, significantly, depend on power from oil and gas to acquire raw materials and to manufacture, transport, and install them. Large hydropower is certainly dependent on cement manufacture, a large source of greenhouse emissions, and on trucks, cranes, mineral extraction, and steel making—all of which are powered by fossil fuels. Likewise, the metals processing and assembly machinery for solar and wind generators now run on oil and gas, although it is possible that large parts of the manufacturing process could be converted to renewable electricity. Agriculture, too, is currently greatly dependent on fossil fuel, especially for the manufacture of fertilizer, the operation of machinery, and transport. One of the sad symptoms of near-total dependence on petroleum is that, apparently, very few in the realm of national policy and industrial planning have given much thought to a life with little oil.[13]

As it was before oil was discovered, coal will be available, and there are many more years of reserves. Russia, China, India, and Australia all have huge coal deposits, and the United States has the largest proven coal reserves of any nation, about 270 billion tons (though it imports more than 30 million tons a year of low-sulfur coal from South America and Indonesia). The official prediction is U.S. reserves will last 150 to 200 years. World reserves are somewhere near 1 trillion tons. However, coal provides 25 percent of world energy—including 40 percent of the electricity—and demand is rising, especially in China, which burned a third of what the world used in 2005.

The bad news is, burning coal contributes 80 percent more heat-trapping gas emissions than oil (for an equivalent energy value). Burning coal for electricity spews out more than 18 percent of the U.S. annual output of nitrogen oxides and 55 percent of sulfur dioxides, 37 percent of CO_2, most of the particulate air pollution, and 66 percent of the mercury. And coal smoke, though scarcely visible as it leaves the stack, is a deadly brew that contaminates and sickens far downwind. Between 24,000 and 30,000 Americans die each year from pollution generated by coal-fired power plants. Ecosystems, too, can be devastated by mining, especially such indiscriminate practices as mountaintop removal. Nevertheless, engineers in the Department of Energy and at various energy companies plan to build more than one hundred more coal-fired plants in the United States (although many of the proposals are being challenged).[14]

Indications are, moreover, that only a few of these plants will be outfitted with the most efficient available technology, known as integrated gasification combined cycle (IGCC) technology. This converts pulverized coal into a high-hydrogen gas that is then used to fuel a gas turbine which also makes steam for conventional power generation. This technology not only emits less carbon dioxide per kilowatt than conventional coal-fueled power plants but also allows for easier CO_2 capture, storage, and sequestration (CSS)—the permanent isolation of carbon away from the atmosphere by physical, chemical, or biological means. One method involves pumping pressurized CO_2 into depleted oil wells, where it would either be contained or used to coax out residual oil. The first IGCC coal power plant with underground CSS is still being developed. It is very possible that the process will not work or will face enormous cost and liability burdens, but the U.S. government and several power companies have test programs.[15]

Although turning coal into gas is a well-known technology and the IGCC recovery of heat using more than one thermodynamic process is common, only a few large coal power plants use them together. Only two are operating in the United States. These plants cost about 20 percent more to build, and power companies have been betting that they wouldn't be required to sharply limit CO_2 emissions. But regulatory laws are imminent, and firms are looking at IGCC as well as other technologies. A single 1,000-megawatt coal-fired power plant emits as much CO_2 as about 1.5 million cars—7 million tons—a year. More than six hundred large coal plants now produce 50 percent of American electricity. China and India, meanwhile, are adding coal power plants at an astounding rate: 560 are proposed in China and more than 200 in India by 2012. If the proposed 860 new plants are built in the United States, China, and India, they will cough out more than 2.5 billion tons of CO_2 annually—five times the reduction goal of the Kyoto protocol. These plants do more than just foul the air: with their fifty-year life spans, they obstruct the changeover to less polluting plants, since both companies and governments will fight to avoid the costs of shutting them down.[16]

Thus the most serious global warming threat is from coal. This fact, as well as the world's dependence on oil and the overall unsustainability of our energy use, reveals the stark truth that climate change is, more than anything else, an energy problem the size of our civilization. We need to stop using nineteenth-century fuels in the twenty-first century.

Our huge need for fossil fuels really began barely one hundred years ago. The way we use the land has a much older history, but the percentage of the Earth given over to human use has increased with population and the harnessing of energy. This has encouraged sprawling cities, deforestation, single-crop agriculture, and billions of domestic grazing animals, all of which also create greenhouse gases. Livestock grazing, feeding, and processing, according to a 2006 report by the Food and Agriculture Organization of the UN, not only uses a third of arable land but also puts out 18 percent of all human greenhouse emissions. Producing meat and milk and other animal products accounts for 37 percent of our methane and 65 percent of our nitrous oxide output, with a total CO_2-equivalent greenhouse-gas level that is higher than that generated by the world's vehicles and transportation. Like the use of coal and petroleum, land use changes and agriculture are inseparable from our economic and social systems. Peak oil arguments are really moot. With severe climate disruption looming, finding a way to change or adapt to put out less polluting gas is a crucial new task for civilization.[17]

The solution may lie in recognizing the broad damage caused, far beyond a narrow definition of environment. Climatologist Stephen Schneider of Stanford University and colleagues have introduced what they call the "Five Numeraires" of the impacts of climate change. They were reacting to the all-too-common tendency for costs of social and political decisions (in this case, actions mitigating climate change) to be couched only in market terms—such as the "losses or gains to GDP" statement often used when national programs are debated. Yet gross domestic product counts only the total value of final, new goods and services produced in a year. "These models are only for the rich," Schneider has written. "Equity isn't an issue[;] . . . nature doesn't have a vote. Economists looking at standard cost-benefit models treasure only what is traded." That is, GDP does not consider the poor health that requires the medical expenditures that *are* factored into GDP; it does not include the loss of wetlands

Cooling stacks of the 2,900 megawatt Amos coal-fired power plant near Charleston, West Virginia, looming over a neighborhood in Poca, across the Kanawa River. The visible steam is benign compared to the nearly invisible but harmful emissions from the stacks, which contain carbon dioxide, particulates, sulfur, and mercury. This plant's CO_2 output has hit record-high levels. However, the company that owns the plant has begun to experiment with carbon sequestration at two of its eighty other power facilities. (SEPTEMBER 2005)

and their flood protections when a new mall is built, though its construction costs *are* counted. It counts the costs of cleaning up an oil spill but not the value of ecosystems damaged by the oil.[18]

Schneider and colleagues, recognizing the limitation in economists' tallies of global warming costs, applied some corrections. They start with the market costs—the traditional GDP reckoning—but then go on to include loss of human lives, loss of biodiversity, income redistribution, and quality of life (loss of heritage sites, forced migration, disturbed cultural amenities, etc.). This system acknowledges that while developing nations may not experience as much dollar loss from climate change as the developed world, they stand to lose much more in human lives, natural resources, and way of life. Looked at in this way, the present and foreseeable costs of climate disruptions are vast and span the interconnected spectrum of human life and endeavor.[19]

If we focus too narrowly on the economic balance sheet of energy use and climate change, we miss the lessons of science and history. In *Collapse*, geographer Jared Diamond lists climate change as one of twelve ways in which past societies have undermined themselves. Nations can be blind to the threat for many reasons: the lure of power, outdated beliefs, the incremental and invisible nature of the changes, and the conflict over resources between individuals and the common good. Not often mentioned is the fact that for the few millennia that humans have farmed, tended livestock, gathered in cities, and developed technology, the Earth's temperature has been relatively stable. "Yet today," wrote Thomas Lovejoy and Lee Hannah recently in the *International Herald Tribune*, "despite all the signals from nature, we are failing to come to grips with the fact that we are changing the climate ourselves."[20]

THE ROAD BEYOND KYOTO

We are changing the climate. This is a powerful truth, considering climate's controlling role in creating the habitat of every living thing and its influence over the course of human events. It is no wonder that the word *climate* has evolved in our language to mean also a set of attitudes or conditions in human affairs. Influence and effect flow both ways between the climate of the planet and that of our social and political realms.

Just as elections, revolutions, and social upheavals initiate change in society that can last for many years, greenhouse gases remain aloft for hundreds of years, with effects far into the future. Yet as we increasingly see the harmful results of energy and landscape management decisions, we must also see that different choices can create change in the opposite direction. Every level of community, business, national governance, and international relations must be involved in this, and most changes are going to be neither easy nor quick. Some actions are obvious and will pay for themselves; others will rely on new ideas and new technology and will change how money is spent. However, many other benefits will accrue as we shift not just our energy sources but also our habits and assumptions. The shape of this change is visible already. Many people, companies, and nations know how to take action and are already succeeding in reducing their impact on the atmosphere.

One source of optimism is the fact that more than 162 nations and other parties to the 1992 Framework Convention on Climate Change have signed the Kyoto Protocol on reducing greenhouse gases, which took effect on 16 February 2005. The protocol requires modest reductions in greenhouse gas

THE UN CLIMATE TREATY

FRAMEWORK AND PROTOCOL

In the 1960s, the astronauts of the Apollo Project gave us the first color photographs of Earth from space. The vision of our lovely, fragile planet set against the darkness of space underscored our connectedness as we began to realize the global reach of pollution and greenhouse gases. The atmospheric temperature had been rising throughout the early part of the century but had cooled slightly beginning in the World War II years. Even though some scientists were suggesting that a very gradual slide into a new ice age had begun, contrary evidence was mounting. By the late 1950s, oceanographer Roger Revelle had shown that man-made CO_2 was not going to be fully absorbed by the oceans, and Charles Keeling had actually measured a yearly increase of the gas in the air. The year 1970 saw not only the establishment of the U.S. National Oceanic and Atmospheric Agency (NOAA) but also the first Earth Day. During these years, many nations passed laws against pollution, and international meetings were organized about weather and climate. UN concern for the environment was focused by the World Environment Summit in Stockholm, Sweden, and establishment of the UN Environment Programme (UNEP) in 1972.

By the time atmospheric temperatures had begun to rise again in the 1970s, damage to the ozone layer was discovered. Modern chemicals used as refrigerants, propellants, and cleaners were implicated—once again showing the planetary reach of our technology. Individual nations, including the United States, banned some of the chemicals, and international

talks on this issue culminated in the 1987 Montreal Protocol on protection of the ozone layer.

Worldwide temperature increases in the 1980s raised concern among some world leaders. Discussions began about creating a world treaty on climate change. In 1988, the World Meteorological Organization and UNEP set up the Intergovernmental Panel on Climate Change as an advisory group. The IPCC does not carry out research; rather, it has enlisted more than 2,500 qualified scientists (who volunteer their time) from 130 nations to analyze peer-reviewed science reports and assess what we know about the climate system and the rate and impacts of climate change. The group assembles the results and arrives at a consensus about trends and causes. It has issued four reports—in 1990, 1995, 2001, and 2007—and each report has revealed an ever-richer knowledge of the climate system and human-caused global warming.

Following the first IPCC report in 1990, the United Nations held meetings that resulted in the Framework Convention on Climate Change (UNFCCC), also referred to as the UN climate treaty. The treaty was presented for signatures during the Rio Earth Summit in 1992 and by now has been ratified by 189 nations. It was signed and ratified by the U.S. Senate during the George H. W. Bush administration. As a ratified international treaty it has the same status as a federal law.

The convention's objective is "to achieve stabilization of greenhouse gas concentrations in the atmosphere at a low enough level to prevent dangerous anthropogenic interference with the climate system. Such a level should be achieved within a time-frame sufficient to allow ecosystems to adapt naturally to climate change, to

ensure that food production is not threatened and to enable economic development to proceed in a sustainable manner" (Art. 2).

The convention continues with a long list of principles and actions to be followed by the nations that are party to it, including:

• *The Parties should protect the climate system for the benefit of present and future generations of humankind, on the basis of equity. . . . Accordingly, the developed country Parties should take the lead in combating climate change and the adverse effects thereof. . . .* ("Principles," Art. 3.1)

• *The Parties should take precautionary measures to anticipate, prevent or minimize the causes of climate change and mitigate its adverse effects. . . .* (Art. 3.3)

• *The Parties should cooperate to promote a supportive and open international economic system that would lead to sustainable economic growth and development in all Parties, particularly developing country Parties, thus enabling them better to address the problems of climate change. . . .* (Art. 3.5)

Signatories of the Convention will also

• *Promote and cooperate in the development, application and diffusion, including transfer, of technologies, practices and processes that control, reduce or prevent anthropogenic emissions of greenhouse gases. . . .* ("Commitments," Art 4.1c)

• *Cooperate in preparing for adaptation to the impacts of climate change. . . .* (Art. 4.1e)

• *Each of [the developed nations] shall adopt national policies and take corresponding measures on the mitigation of climate change, by limiting its anthropogenic emissions of greenhouse gases. . . .* (Art. 4.2a)

Because the UNFCCC has no deadlines or procedures for achieving these objectives, the Kyoto Protocol to the Convention was negotiated in 1997 to set specific goals for greenhouse gas reductions. However, the protocol was rejected by the U.S. Senate during the Clinton administration and denounced by President George W. Bush. The Kyoto Protocol reached a working membership of 159 nations in February 2005, without the United States. Australia joined in 2007.

The Kyoto Protocol requires modest reductions in greenhouse gas emissions, to be instituted first by the most developed nations, which would lead to emissions about 5 percent below the 1990 level by 2012. Opponents within the Bush administration reportedly regarded the protocol as an economic attack, "an elaborate, predatory trade strategy." An Australian minister called it "an environmental disaster." A primary objection was the failure to require China and India to make emissions cuts.

George Bush's refusal to go forward with Kyoto created an international rift between the United States and other nations, exacerbated by the fact that the United States is responsible for one quarter of fossil fuel emissions worldwide. At the first meeting of members of the protocol in Montreal in December 2005, the United States attended as an outsider, yet its delegation tried to get in the way of implementation and talks on future emission reductions. Nevertheless, the meeting ended with the protocol fully operational and nations promising to meet to decide on further controls beyond 2012, to work toward the deep cuts in yearly greenhouse emissions that will be needed to level off increases in warming. Little progress was made at the second meeting, in Nairobi in 2006, but funds were organized to help less-developed nations adapt to climate change effects. In 2007, under pressure from the new IPCC report, nations overcame resistance from the United States, China, and other countries and agreed to forge emission limits to take effect when the Kyoto Protocol expires in 2012.

Sources: The most complete history of climate science and its connection to policy is Spencer Weart's *The Discovery of Global Warming* (Cambridge, MA: Harvard University Press, 2003), together with his updated website at www.aip .org/history/climate. The full Convention and Protocol may be read at http://unfccc.int/ essential_background/convention/background/ items/2853.php. "Predatory trade strategy" quote appeared in "Blair Turns Up Global Warming Heat," a column by Robert Novak, 13 June 2005 (available at www.findarticles.com/p/articles/ mi_qn4155/is_20050613/ai_n14717327); Ian Campbell, Australian Minister of Environment and Heritage, is quoted in an interview at the IMAC1 convention, Geelong, Australia, 24 Oct. 2005. ✦

emissions by 2012 (to an average 5 percent below the level in 1990), first by the most developed nations. The United States, however, alone among industrialized nations, has not ratified the Kyoto Protocol. This obstinacy ripped a hole in international relations and shredded opinion of the United States, because everyone knows who is responsible for a quarter of the problem. Mechanisms of the protocol are in effect and nations are promising to meet regularly to discuss further controls beyond 2012. So far those talks have not created an ongoing plan. The United States and Australia announced in 2005 a separate "Asia-Pacific Partnership on Clean Development and Climate" together with China, India, Japan, and South Korea. In the United States at least, funding for this effort is paltry; moreover, it is unclear how this extra-Kyoto pact will work and what voluntary emission reductions may come of it.[21]

The Kyoto agreement is but a baby step in reducing world carbon use and output. It covers less than 30 percent of world emissions and imposes puny limits on CO_2. However, as time goes by and emissions grow, the reduction based on 1990 levels looms larger and larger for some industrialized nations. Canada is 28 percent over its 1990 levels, and Japan, too, is emitting far too much. The United States continues to pump out about 16 percent more greenhouse gas than it did in 1990. However, Germany is emitting significantly less than it was in 1990, aided by a shutdown of inefficient coal-burning plants in the former German Democratic Republic. The United Kingdom is also within its Kyoto target, as are France and Sweden. China, which as a developing nation under the climate treaty has no mandatory Kyoto limits, is almost 50 percent above 1990 levels.[22]

Nevertheless, the Kyoto Protocol signals an international awareness of the problem, and it has instituted a procedure with which to move forward. If the political and industrial worlds are to heed the scientific warnings and try to head off the dire predictions, then cooperation, commitment, and a plan of attack are essential. The protocol has encouraged a huge amount of study, planning, collaboration, and concern in the private sector alone. Many countries in the European Union and elsewhere have plans to reduce emissions far beyond the Kyoto levels.

Even as nations were ratifying the Kyoto agreement, the rate of increase of methane was slowing, and emission of chlorofluorocarbons (CFCs), another set of greenhouse gases, was falling sharply. Indeed, the Montreal Protocol of 1987, which limited the release of the ozone-depleting CFCs internationally, is a model of cooperation on a worldwide problem and was an inspiration for Kyoto. Fossil fuel emissions in the United States are now increasing at around 1 percent per year, down from over 4 percent through the 1970s. The oil crisis of that decade led to gasoline mileage standards, reduced industrial waste and emissions, and much more efficient household appliances. Since 1974 Americans have saved about 200 billion kilowatt-hours per year from increased refrigerator efficiency alone. Those new refrigerators cost a lot less, too. As a result of these changes, the United States uses 47 percent less energy per dollar of economic output than thirty years ago (this ratio—energy use per unit of GDP—is called energy intensity). Today, the world is using less fuel than was predicted at the height of the 1970s oil crisis. This should encourage those who think we have no options. Efficiency and cooperation— and higher oil prices—work.[23] Although some economists and politicians argue that the improvements we have instigated mean no further cuts are required, that the market will drive a slowdown in emissions, this is clearly not true. Recently the reductions in total world energy intensity have stopped, according to the Global Carbon Project. More CO_2 was put into the air in 2007, both absolutely and in parts per million, than ever before, and total CO_2 emissions are increasing four times faster than they were during the 1990s.[24]

Holding temperature change this century to less than an additional 2°F (1.2°C) will require, according to some estimates, total global CO_2 output to drop by more than half before 2050 and then continue to stay at or below that level despite population growth and development. Because most of the CO_2 now comes from a handful of nations, they must bear most of the burden (meaning up to 90 percent cuts, by some estimates); for humanitarian reasons room needs to be allowed for lesser-developed countries to bring up their standards of living. The longer we wait, the more expensive it will become and the more severe the measures needed to force what could now be done voluntarily and profitably. The top-heavy task under Kyoto of ranking the world's nations according to their population, economy, and carbon emissions and deciding which ones should move first to cut greenhouse gases is not the only model for cooperation. One of the most highly developed models is "contraction and convergence," which leads from egalitarian ideals by way of science. The best estimate of the amount of greenhouse gas in a stable atmosphere well short of catastrophic change would be the target, and nations would move toward it (contraction) based on an eventual equal distribution of emissions per person (convergence). Proponents see this equality as the only humane way of apportioning "use" of the atmosphere, which has no boundaries and supports everyone. It could be run like a global cap-and-trade scheme, in which a limit (the cap) would be set and participants would cooperate through shared innovation and market trading to meet the overall goal. For example, early in the process poor nations with many people but few emissions could sell some of their allowances to raise capital for an industrial base that would increase standards of living, while industrialized countries would work to invent ways to lower their per-capita emissions to within the limit. The Global Commons Institute has been bringing this idea before international climate meetings since 1990.[25]

However they come about, new controls on greenhouse gas emissions will have to be made across the entire spectrum of energy production and use. The goal is to still enjoy abundant energy but to get it from different sources.

The remainder of this chapter will provide an overview of the energy sources available right now and discuss some of the decisions that must be made about them in the near future. An intelligent and fast-acting program for moving toward the best energy sources will have to involve equitable costs for carbon emissions and fair limits on greenhouse gas emissions; a level economic and legal playing field for all energy sources, purveyors, and users; and an open marketplace in which pollution level, safety, siting, and price will select the mix of sources. The fact of global warming is that we are going to need some combination of all kinds of clean energy.[26]

FUEL: FROM FOSSIL TO FUTURE

One energy source that is already in use but is controversial and fraught with dangers of its own is nuclear power. Its great benefit is that, because it uses the heat of uranium fission, it does not directly create greenhouse gases. As an MIT study put it in 2003, "The nuclear option should be retained precisely because it is an important carbon-free source of power." Climatologist John Christy told me that "the only path to CO_2 reductions that would have an impact . . . is a massive nuclear power construction era." Much of the public and some energy analysts, including those at MIT, summarize the drawbacks in four phrases: overwhelming cost, Yucca Mountain, nuclear proliferation, and Chernobyl.[27]

Some 440 nuclear plants are in operation worldwide, generating about 16 percent of world electrical power. Although the United States leads in the overall amount of electricity gained from nuclear technology, France is the most committed to it, getting more than three-fourths of its electric power from nuclear plants. Lithuania, Slovakia, and Belgium also generate more than half their power from nuclear plants. The United States, in contrast, garners 20 percent of its electricity from 103 nuclear installations in thirty-one states. Typically, these plants are water-cooled. They are also very inefficient, getting the energy from only about 5 percent of the enriched uranium fuel and creating 100 tons of spent fuel a year per 1,000-megawatt plant. Although nuclear is touted as a "zero-carbon" power source, where we get the uranium and what happens after the power is generated complicate this assertion. Mining, transporting, and processing uranium all lead to CO_2 emissions. The heat given off by these plants and their need for coolant water can be problematic as well. During the European heat wave of 2003, at a time of great electrical demand, some French plants had to shut down for lack of water.[28]

Waste is a huge issue with nuclear power. When the fuel in a water-cooled reactor reaches the end of its useful life, it is still radioactive. A tiny part of it is plutonium, which can be made into bombs and remains radioactive for thousands of years. It is no secret that the world has no permanent facilities for this waste. In the United States alone, which is struggling to establish a depository at Yucca Mountain, Nevada, 58,000 tons (53,000 mt) are already being held in temporary storage at nuclear plants throughout the country. The interconnection between peaceful nuclear technology and military uses raises fears of proliferation and of the nuclear material falling into the hands of terrorists. And although the chances of another Chernobyl happening may be slight, "incidents" still do occur; the specter of a meltdown or malfunction like that at Three Mile Island, Pennsylvania, in 1979 continues to haunt the public.[29]

Operating costs are also very high, and private capital cannot manage without government subsidies and protection against losses and liability worth many billions of dollars. In 2005 the Rocky Mountain Institute, an energy think tank, surveyed costs and savings of various energy sources and found that even with market and tax assumptions that favor nuclear plants, the cost per kilowatt is higher for nuclear than for wind power and much higher than for cogeneration or improved efficiency in existing power sources. Nuclear plants also take longer to build and test than other energy-producing facilities. Primarily because of the cost and safety issues, no plants have been built in the United States for three decades, though about thirty are currently under construction in Europe and Asia. The future possibilities for nuclear power may brighten if new designs can come into play for plants that are less expensive to build, more efficient to operate, safer, and longer-lived. Advocates of another type of reactor, commonly called breeders because they can be used to create plutonium, say these machines can also be designed and operated as near-total consumers of the fuel, which would help solve the nuclear waste issue. As one solution to global warming, nuclear power has the advantage that it is in widespread use and has some powerful independent supporters, including James Lovelock. The strikes against it, however, in terms of cost, safety, and security, are daunting, and in the end, without technical and geopolitical breakthroughs, it should not be expanded.[30]

RENEWABLE ENERGY is the direction of the future, even though the technologies involved span the long history of human life. The promise here lies in the sun's energy, which each hour strikes the Earth with more power than humans use in a year. This immense flow drives the weather, the wind, plant growth, and the water cycle. People have been using wind and water power for thousands of years, and the energy content of plants and direct radiation from the sun for even longer. Wood, agricultural waste, and methane from landfills, all of which can be burned, are older forms of bioenergy. Modern inventions include solar photovoltaic panels, wind turbines, and new methods of producing ethanol. Hydropower, solar thermal generation, biodiesel fuels, wave and tidal generators, solar water heating, and ocean thermal energy conversion are other renewables. (Cogeneration, the efficient use of excess heat or waste from an industrial process to generate electricity or heat buildings, is sometimes included in this category, but because it is part of a process that generates CO_2 it should be classed as a more efficient use of current power.) Renewables are the fastest-growing segment of the energy sector, according to an Earthscan report, and accounted for 20 percent of worldwide financial investment in energy in 2004. Still, they provide only about 13 percent of total global energy use, and most of that is from traditional wood fires and hydroelectric dams.[31]

For electrical generation using renewable energy sources, most attention has been focused on wind and solar power, either one of which alone could supply more than enough electrical power for the whole world. Generation by wind is the world's fastest growing energy source, expanding some years by more than 30 percent, with a capacity exceeding 94,000 megawatts. Europe has taken the lead: Denmark has installed wind turbines that can supply almost 20 percent of its energy needs, 3,000 megawatts from more than 6,000 turbines; and Germany has 18,000, generating 20,600 megawatts, the biggest producer in the world. German and Danish machines dominate the world market, and wind turbine manufacturers employ tens of thousands. In those nations and in the United Kingdom, however, critics have wondered just how much power is actually entering the national grid, saying that wind generators are inefficient relative to their cost. Germany's energy agency chief was quoted as saying that more energy-efficient buildings would be much cheaper than wind power per ton of CO_2 saved; still, he said, everything that reduces emissions is needed. He seems to have it right: Germany is trying to stay on track to meet the Kyoto requirement while proceeding to decommission all nineteen of its nuclear power plants by 2020. Its program includes energy-saving targets for homes, businesses, and transportation; a "100,000 roofs" plan for solar; a drive to upgrade insulation and other construction features; toll increases for long-distance driving and tax breaks for efficient vehicles; an emissions trading scheme; and deployment of 5-megawatt wind turbines, the world's largest. Germany expects to get half its electricity from renewable energy by midcentury.[32]

In the United States, utilities have been putting up wind farms rapidly, spurred by high interest in clean power and by tax credits. In 2007, wind turbines generating more than 5,000 megawatts were installed, bringing U.S. wind-generating capacity to more than 16,800 megawatts. As turbines have been erected, wind energy costs have dropped, such that now wind power is competitive in many markets with conventional power (which is also subsidized). Nevertheless, the technology is not without its issues. For example, there is much scientific and public concern about bird and bat kills by the wind turbine rotors. Much of this controversy is driven by the disastrous placement of early wind turbines at Altamont Pass, California, which were so low and tightly arrayed that they made a killing fence for migrating raptors. Some of these older towers, which look more like oil derricks or farm windmills, are at long last being dismantled. Modern turbines are much higher and more widely spaced, but they should not be developed in scenic locations or important habitats for birds and other flying creatures. More than two thousand dead bats were counted in each of two surveys at the Mountaineer wind project in West Virginia, for example. Opposition to wind installations is also fanned by lack of regulations and expertise in local jurisdictions, where most of the permitting takes place. In the United States, if not elsewhere, wind development still needs sensitive national siting and study standards to prevent conflicts with local and natural values.[33]

It may be that wind generators installed offshore in deep water will be a way to increase output, especially because wind is more constant over the ocean. In Britain as well as the United States, proposed near-shore sitings face opposition based on possible interference with shipping, fishing, and

The roof of the Moscone Convention Center in San Francisco, California, is covered by photovoltaic cells, which cool and protect the roof as well as generate enough energy to power a thousand homes. The 60,000-square-foot installation is the largest city-owned photovoltaic solar array in the nation. (JUNE 2005)

overwater views. Another issue is that the best locations for wind turbines will probably not be on existing power grids, requiring large outlays for transmission lines (which have their own cost and placement considerations). On a small scale, wind generators are also a choice for individual homes, businesses, and neighborhoods.

Solar Power includes photovoltaic cells (PVs—semiconductor devices that convert sunlight into electric current), solar thermal (in which sunlight is focused to heat a fluid that drives a turbine), and solar hot water heating. Some analysts believe that solar power has the greatest potential among renewable energy sources. The installation of PVs to connect with an existing power grid, as on many new buildings and retrofits, is particularly fast-growing. Many new commercial buildings are designed with integral PV panels; on one new medical building in Portland, Oregon, every south-facing window awning not only will provide shade but will also be a power source. Incorporating many other ecological features, this high-rise will use 60 percent less energy than the building code specifies. One of the first companies in the world to retrofit a large office building was CIS in Manchester, England, which reclad three sides of its twenty-five-story building's service tower in PV solar panels. About 5,000 modules replaced the previous exterior surface tiles, which had been falling off.[34]

To get much of our grid power from solar in the short term, however, we are going to have to spread the photovoltaic cells around miles and miles of landscape. Updating a calculation first published in *Science* in 2002, a *National Geographic* article figured "it would take about 10,000 square miles (30,000 sq km) of solar panels—an area bigger than Vermont—to satisfy all of the United States' electricity needs. But the land requirement sounds more daunting than it is. . . . All those panels could fit on less than a quarter of the roof and pavement space in cities and suburbs." Every city has acres and acres of parking lots and commercial rooftops that should become solar energy installations, which, along with panels on residential roofs, could generate about three-fourths of U.S. electrical capacity, according to another study. Pioneers in installing rooftop solar panels include Fala Direct Marketing on Long Island, which installed PVs producing 1.1 megawatts over 100,000 square feet (930 sq m) of rooftop in 2002, and FedEx, which covered its Oakland Airport facility with a solar field nearly as large in 2005. They will not be leaders for long: Google plans to generate 1.6 megawatts from the roofs of its Mountain View, California, headquarters. Germany has the three largest PV arrays of up to 5 megawatts each, and one twice as big is going to be installed in Spain with funding from General Electric.[35]

The inventor Stan Ovshinsky is perfecting thin film PVs that are flexible and can be cut to size. This and other coming innovations should soon make solar electric surfaces commonplace on the decorative panels and roof tiles of tens of thousands of buildings, as well as on automobiles, cell phone and computer cases, and even clothing. This direction also makes it more feasible to deploy panels in the remote villages of the world; Kenya already makes much use of standard small panels that can offer electricity to a single house. These developments should lower the price for PVs, which has been higher than that for other energy sources.[36]

Solar thermal electricity generation is proven and operational in the California desert. Nine arrays focusing sunlight with parabolic mirrors run turbines that deliver 354 megawatts to Southern California's power grid. The next generation of thermal generators may be a 4,500-acre solar project northeast

Kramer Junction solar plant, near Barstow, California, is part of a complex that provides 354 megawatts of electricity and is believed to be the largest operating solar power station in the world. This plant uses parabolic reflectors to concentrate sunlight on a fluid that flows through tubes to turn a conventional turbine generator. (JUNE 2005)

of Los Angeles, using Stirling engines now undergoing testing at Sandia National Lab. Parabolic bowls focus the sun's heat on the self-contained cylinders in which hydrogen expands and contracts, moving a piston. This technology may be more efficient than other solar generators, and according to developers this single installation could generate 500 megawatts of electricity, the most of any single solar site in the world. The ecological effects of covering large areas of landscape with solar collectors of any kind could be severe, so siting needs careful study (another reason to consider already committed areas like roofs). Solar water heating is in worldwide use, and once was common in sunny parts of the United States. It's now resurging, but almost all the market is in China, which has 60 percent of the world collector area and in 2004 accounted for 80 percent of all new installations.[37]

What is especially missing from the future of wind and solar energy generation is efficient ways to store energy against times of calm and dark. Batteries may not be the only answer. Excess energy could feed into giant capacitors and be used to compress air or pump water uphill into reservoirs that could later turn turbines. Solar thermal plants can be designed to store hot fluids for use at night. Chemist Daniel Nocera's lab at MIT has found an efficient way for renewable electricity to generate hydrogen for fuel cells from water, by emulating photosynthesis. Until thousands of renewable installations are on line or there is robust storage technology, coal or natural gas power plants may still be needed on many power grids for their constancy.[38]

THE AMOUNT OF POWER in the natural movement of water makes this old resource very attractive for new technologies to make electricity, assuming the sites chosen avoid conflicts with river and coastal values, water animals, wilderness, and shipping. Several approaches are available. Wave and tidal generator studies and tests are ongoing, especially in the United Kingdom and northern Europe, and others are planned in the Pacific Ocean off Oregon. In New York's East River a demonstration project is being installed. The promise of ocean thermal energy conversion stems from the temperature difference between the surface and deepest waters, but the process of tapping into that difference is very inefficient so far.[39]

And then there is more traditional hydropower, which usually means huge dams—a technology that continues to be used despite enormous costs and great damage to the environment. The water in the reservoirs may be renewable, but the values of the river certainly are not, at least on a human time scale. Three Gorges Dam on the Yangtze in China, for example—which may generate as much as 18 gigawatts of energy, equal to that of three or four giant coal power complexes—will cost more than $25 billion (the Chinese estimate, though other estimates range as high as $100 billion). It is also displacing millions of people, contributing to the loss of river dolphins, paddlefish, and cranes, and starving the crucial river delta near Shanghai of silt. Hydropower supplies about 15 percent of world electrical power, and in North America it remains prominent in the West. Small hydropower is more often used in places that are off the power grid, serving small towns or businesses.[40]

Finally, in twenty nations where geysers and volcanic steam vents spring forth, geothermal electrical generating capacity is hard at work—producing about 8,000 megawatts, all told (though only a fraction of the total potential is currently being exploited, according to the Renewable Energy Policy

Iceland's Svartwengi geothermal power station, which uses steam to make electricity for Reykjavik. Geothermal energy also warms the water of the famous Blue Lagoon thermal springs. Geothermal power for human use is limited to places where volcanic heat is safely accessible, but it is a huge potential source for Africa's Rift Valley. (MAY 2001)

Project). California, the Philippines, Italy, Mexico, and Indonesia are leaders in the amount of energy produced, but only Iceland is heavily dependent on geothermal. This power is a natural choice in Africa's Great Rift Valley, but so far only one utility, in Kenya, is tapping this enormous volcanic resource that could serve so many under-electrified nations.[41]

BIOFUELS HARK BACK to the distant past and the harnessing of fire by humans. Wood, charcoal, peat, grass, dung, and other organic material, now called biomass, are still major energy sources: 50 to 60 percent of the energy in developing countries of Asia and up to 90 percent in Africa comes from wood or plant material. According to the World Bank, a third of the world cooks with wood. Because the local air pollution from this method is so unhealthy, especially indoors, there is a push to provide cleaner-burning stoves and alternatives like solar cookers.[42]

Ethanol is also an ancient substance, used as medicine, intoxicant, and for ceremonial purposes for thousands of years. Today, ethanol for motor vehicles is distilled from soybeans, corn and maize, rapeseed, palm oil, sugarcane, wheat, and jatropha (physic nut, a shrub extensively planted in India and arid areas). Brazil is nearing energy independence for its transportation sector by augmenting its petroleum supplies with ethanol made from sugarcane. More than 340 Brazilian sugar mills and distilleries made nearly half the world total of ethanol, about 4.8 billion gallons (18 billion l), in 2005. Car makers there have responded with engines that can accept any mixture of gasoline and ethanol. Because its cane is cut by hand, not machine, and because it lacks an extended fast highway network, Brazil may not be a model for the United States to emulate.[43]

Nonetheless, the use of ethanol as a motor fuel additive is growing in the United States, with production up to 6.5 billion gallons in 2007, 30 percent more than the year before. More flexible fuel vehicles are being sold that can run on a mix of 85 percent ethanol/15 percent gasoline, called E85, but general

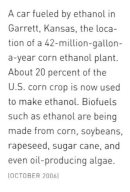

A car fueled by ethanol in Garrett, Kansas, the location of a 42-million-gallon-a-year corn ethanol plant. About 20 percent of the U.S. corn crop is now used to make ethanol. Biofuels such as ethanol are being made from corn, soybeans, rapeseed, sugar cane, and even oil-producing algae. (OCTOBER 2006)

acceptance awaits the fuel's distribution to many more of the country's 126,000 gas stations. Ethanol has been known in the United States for many decades as a fuel additive (some drivers will remember "gasohol" from the 1970s), but the major oil companies opted for the cheaper lead and then MTBE (methyl tert-butyl ether) to increase power, with disastrous results for public health. Ethanol has few health issues identified so far, and although it does result in a slight loss of fuel efficiency, the important fact for climate change is that even mixed at only 10 percent with gasoline, it leads to a 19 percent reduction in greenhouse gas emissions. At the same time, corn and soybean growing as now practiced in the United States and elsewhere results in great loss of topsoil, pollution from pesticides, and fertilizer runoff that is poisoning rivers and forming an oxygen-depleted zone in the Gulf of Mexico. An emerging technology, however, promises to make ethanol from a wide variety of crops as well as agricultural and timber waste, using not just starches and sugars, as with corn, but the entire plant. The product is cellulosic ethanol, which, according to a review by the Natural Resources Defense Council, "delivers profoundly more renewable energy than corn ethanol [and] . . . promises to consume less petroleum, produce fewer greenhouse gases, and require less land compared to corn ethanol." Another new technology is being piloted that feeds oil-producing algae with industrial exhaust rich in carbon dioxide and nitrogen oxide; the process cleans up the exhaust by stripping it of up to 40 percent of the carbon, and the algae themselves can be harvested and converted to biofuel.[44]

Forward-thinking chemists, engineers, agriculturalists, and environmental scientists are beginning to experiment with loop systems that would generate methane from farm waste for heating and power and also fire the biorefinery that is producing ethanol. A further development may be to combine biotechnology, chemistry, and genetics on the farm. Such a system could grow certain plants specifically to be refined into not only fuel but also raw materials for plastics, fabrics, and chemicals that now are refined from petroleum. The basics of bioplastic have been known since the 1930s, when Henry Ford used some on his cars. Cargill has been making a biodegradable plastic for food containers, and DuPont is in the game as well. A wide range of products can be made for about 30 percent less energy per pound. Biodiesel, made from most vegetable oils and animal fats, including waste cooking oil, and even from some types of algae, is another growing market, one that recalls Rudolph Diesel's demonstration engine in 1900, which ran on peanut oil. Biodiesel is biodegradable and nontoxic and can be used in any diesel engine. It can reduce carbon dioxide emissions by up to three quarters, has fewer particulates, and emits no sulfur. This last feature makes it a particularly attractive means of meeting U.S. federal antisulfur regulations and cutting acid rain in urban regions. Use in truck fleets has grown due to tax breaks. As of 2002, the majority of biodiesel worldwide was made from rapeseed oil, with only a small amount made from sunflower seeds and soybeans. In tropical parts of the world forests are being logged to make room for oil palms. Indonesian production is twice what it was in 1999; in Borneo twenty-seven times more land is planted with palms today than in the mid-1980s. With the concern that deforestation will accelerate to fuel diesel vehicles, the European Parliament is considering a ban on palm biofuel, and nongovernmental organizations are stepping up campaigns against it.[45]

If biofuels must replace gasoline and other fossil fuels to further reduce greenhouse gases, they must themselves be as clean burning as possible. The energy input in the form of agrochemicals and fuel for farm equipment, transport, and processing needs to be closely considered, as does the issue of how climate change itself will affect growing areas. Using coal to power ethanol generators, for example, is counterproductive, because it emits a third more CO_2 than natural gas and must be mined and transported at an economic and atmospheric cost; local gas generated from waste would be a better fuel. The trade-offs and costs of the entire process of making gasoline and its alternatives, called life-cycle accounting, have occupied a number of researchers and economists in recent years. Their studies posit the energy return on investment (EROI)—the ratio of the energy in fuel compared to the fossil fuel used to produce it. The net U.S. EROI for gasoline, from exploration to gas station, ranges from 6 to 10, indicating a positive return on investment. The goal of creating new liquid and gas fuels was stated by University of Minnesota ecologist Jason Hill in a 2006 paper: "To be a viable alternative, a biofuel should provide a net energy gain, have environmental benefits, be economically competitive, and be producible in large quantities without reducing food supplies." Everything that goes into growing a crop and turning it into fuel requires energy input and gives off CO_2, except the plants themselves, which are soaking up solar radiation and some of that CO_2. The energy accounting begins with drilling and refining the fuel for tractors and proceeds through planting, treating the crop with fertilizer and pesticides, irrigating, transporting and feeding the field laborers, harvesting, storing, and, of course, refining the crop into biofuel. The focus of these studies in the United States is corn, because almost all American ethanol is made from corn kernels (about 27 percent of the harvest in 2007, up from 14 percent in 2004). In most published research on the subject, including Hill's, ethanol from corn yields at least 25 percent more energy than is invested (EROI of 1.25) and a net reduction in greenhouse pollution compared to the fossil fuels it replaces. Biodiesel made from soybeans typically is quite a bit better on both counts. And cellulosic ethanol, analyzed in four studies, had an EROI ranging from about .7 to over 6. The lower figure is from studies by Cornell University ecologist David Pimentel and his colleagues. The way they do the math, "ethanol is not a renewable energy source, is not an economical fuel, and its production and use contribute to air, water, and soil pollution and global warming." Pimentel's research sees much higher costs of the fossil fuel input to the process. Other recent studies show downsides in pressure on other food crops and release of carbon to the atmosphere when valuable land such as wilderness is cleared to grow more and more crops for biofuels.[46]

In the rush to produce biofuels, one thing should be kept in mind: we do not want to be in danger of "eating our seed corn" by using it for ethanol. Nor the soybeans for biodiesel. If the United States chases the dream of energy independence using corn, it will be as futile as trying to drill our way to fossil fuel freedom. The real bottom line was put well by ecologist Hill: "Even dedicating all U.S. corn and soybean production to biofuels would meet only 12 percent of gasoline demand and 6 percent of diesel demand." Ethanol demand is rapidly pushing corn prices higher, affecting food prices. For our sake and the world's, the United States should protect its grain crops, which help feed millions of hungry people. The world must move rapidly to cellulosic ethanol, which many experts think has a high chance of being improved by development and innovation. As ethanol grows in importance, it should be made from plants that can be grown without fertilizer or tilling and processed economically, and that disturb neither crucial natural ecosystems nor rich farmland. In the United States, the Energy Future Coalition is promoting "25 by 25"—the idea that a quarter of U.S. energy will come from agriculture by 2025, along with safe and affordable food and fiber. Bioenergy is also available through the processing of waste oils, fats, plant residue, tree-cutting slash, and other material that is now burned or discarded and not needed to enrich the soil. However, biofuels are not likely to replace much of our gasoline, as profligately as we presently use it.[47]

THE "HYDROGEN ECONOMY" is a catchphrase for a future direction in energy use that will supposedly stave off climate change. Hydrogen, when burned or used in an electric fuel cell, combines with oxygen to make only water. Demonstration projects, such as fuel-cell buses in Europe and a few vehicles in the United States, promote hydrogen as a replacement for gasoline. However, the technology to produce, store, transport, and use it for this purpose is many decades away. Low-carbon sources of energy can be used to make hydrogen, especially at the local point of use, but presently it can't be produced in large quantities without fossil fuels. Physicist John Holdren of Harvard University and the American Association for the Advancement of Science warns that hydrogen in these applications is "merely an energy carrier that, like electricity, may be prized for its convenience, versatility and low environmental impact at the point of end use, but requires the use of a primary energy source for its production." We need long-term research to discover and perfect new technologies, and hydrogen will be among them. The immediacy of climate change, however, requires a focus on the energy savings and sources that we can exploit now.[48]

GREEN CARS AND WILD CITIES

As we look toward low-carbon sources of energy, the biggest issue will be how the energy-intensive segments of our world economy will react. Already many industries, governments, and localities are making progress. In the United States at least, road vehicles require serious examination—and improvement. In many nations low-emission and high-mileage vehicles have been common for many years. Not so in the United States. As federal air and water pollution laws were passed beginning in 1970, automakers were allowed exemptions in emissions and mileage standards—supposedly as a break for farmers and working people with pickups, as well as to help American Motors, which had bought the Jeep brand, a World War II icon. This loophole brought us the Jeep Wagoneer and Chevy Suburban, as automakers abandoned the station wagon. Then followed all manner of sport-utility vehicles—built on a truck chassis and so exempt from passenger car mileage (and safety) standards. Sales took off, truck assembly lines were rapidly converted to make Explorers and Escalades, and car makers have successfully defended the loophole ever since. As a result, from 1990 to 2003 total carbon dioxide emissions rose 20 percent for all cars and light vehicles but 51 percent for SUVs and pickups.[49]

Now, with global warming and high gas prices looming, hybrid vehicles should be flooding the assembly lines. So far, however, only 0.4 percent of all light vehicles in the United States—close to 1 million cars—are equipped with hybrid technology. Even Honda and Toyota, leaders in hybrid design (Honda's Insight and Civic and Toyota's Prius were the first on the market), restrict these engines to a few models. Hybrid SUVs are just beginning to coast into Ford, Mercury, Mazda, and Toyota show-rooms. Luxury hybrids, such as Lexus's GS, are being offered, but they cost a few thousand more and get only 25 miles per gallon in town. Some larger vehicles marketed as hybrids by General Motors are what the Union of Concerned Scientists calls "hollow hybrids," because they improve mileage by only 1–3 miles per gallon. Numbers and mileage matter, according to the Earth Policy Institute, because "if over the next decade we convert the U.S. automobile fleet to gas-electric hybrids with the efficiency of today's Toyota Prius, we could cut our gasoline use in half." The Prius is rated at 60 miles per gallon, but like most vehicles gets about 10 percent less in actual use (the EPA will begin using a more realis-tic fuel economy calculation in 2008).[50]

Yet, perversely, automakers continue to battle necessary advances. Many Californians won't forget the EV1 electric car, which General Motors introduced in 1996—but recalled after 2000—crushing most of the vehicles at the GM desert test track. And today, car makers are fighting in court against in-creased efficiency standards in California—at the same time that they are meeting the stricter require-ments of Canada and Europe. The EU standard will be 43 miles per gallon in 2010.

The American auto fleet averages 21 miles per gallon—lower than in the 1980s. However, the U.S. Congress passed a new mileage standard of 35 miles per gallon as part of the new 2007 energy bill. And although automakers insisted on a phase-in lasting until 2020 and loopholes for certain vehicle types, even a small increase in all vehicle mileage, say 3 miles per gallon, would save one million barrels of oil per day. That would greatly offset the 1.5 million barrels of crude a day that we import from Saudi Ara-bia and the 600,000 barrels from other Persian Gulf nations. Even more gas could be saved and pollu-tion prevented if hybrid cars had plug-in batteries, so that short trips would use no gas at all. In an announcement that raised both hope and skepticism, GM said in late 2006 that it would develop such a vehicle. As for urban dwellers, who use up to a third of auto gas, they should actually be using public transit or traveling by bike or foot or Segway whenever possible.[51]

Unfortunately, the six biggest automakers, including Toyota and Honda, all have huge and rising "carbon burdens," according to Environmental Defense, which computes the total CO_2 emissions of all the cars on the road. The world awaits an international car company to be a hero of this moment: installing hybrid, flex-fuel engines in *all* their models, from pickups to roadsters, and equipping them with efficient air conditioning, tires, and transmissions. Cars can be made lighter and stronger with composite material. All this technology is available right now, and automakers have been using some of these improvements—but chiefly to increase power, not to cut fuel consumption. The 2006 models are, by the way, the heaviest in EPA records. Now, with customers forsaking the SUV and heavy pickup for economical models—hybrid sales are up nearly 40 percent—automakers are belatedly closing truck assembly lines and advertising fuel economy. The market for alternative transport technologies is enor-mous and can only grow, fueled by rising gas prices and knowledge of global warming. After this first step, we all must confront a bigger challenge: bringing to an end the era of sprawl and endless driving.[52]

CHANGING WHAT and how much our vehicles burn is really just the beginning. The automobile has dictated the shape and infrastructure of cities, formed urban society, and helped create the massive inequality in world income and energy use. The modern megalopolis, to quote urban ecologist and planner Richard Register, is not only the largest creation of humanity, it is also "an unsustainable urban system that requires massive networks of streets, freeways, and parking structures to serve congested cities and far flung suburbs." We will not begin to reach the changes in emissions required in the coming years without transforming the city.[53]

In 2006, the year the U.S. population reached 300 million, the world became predominantly urban. Half of the world's 6.5 billion souls now live in town, most not in the megacities with populations exceeding 10 million but in smaller cities. China has ninety cities with a population of more than a million, but Shanghai is the only one with megacity status. Cities occupy just 2 percent of the Earth's land surface, but one estimate is that they eat up three quarters of the world's yearly resource use. London's 7.5 million citizens, according to Fred Pearce in *New Scientist,* have a total ecological "footprint" that is 125 times the area of the city itself, when one tallies up the millions of tons of oxygen, water, food, wood, steel, plastics, and fuel that they consume each year and their massive effluent. Populated areas push outward, aided by the automobile, to cover more and more square miles of countryside and flood the air with heat and greenhouse gases.[54]

Cities, of course, are also the focus of commerce, industry, government, education, and the arts and are engines of huge amounts of income and foreign exchange. The very fact of so many people living side by side has created economies of scale, especially for energy and material distribution, inner-city buildings, and mass transit. Some major cities, notably New York, have used these advantages to create a safe environment and conserve energy; New York runs on half the energy per capita of other major cities, according to Register. The major result of urbanization, however, has been the separation of work from home, shopping, and recreation, along with an increasing dependence on cars to get back and forth. This requires more trucks to serve the sprawl, more roads to carry the traffic, more concrete poured through central cities as multideck freeways and parking structures. The fumes, disruption of neighborhoods, deaths and injuries, illness and waste that go with auto transport have blighted life in many cities. In the late twentieth century in the United States, money was increasingly taken from public transportation and directed into highways and automobile transportation. Register says the entire federal appropriation for the U.S. Amtrak passenger rail system, $1.2 billion in 2008, is the average price of a mere mile and a half of urban interstate freeway. Beijing has closed streets to the traditional bicycles to make room for cars, and it, like many developing cities, is rapidly following the American highway. This is why Register says that improving car mileage is just going to allow us to drive more and "postpone the time when we must redesign the entire city."[55]

By "redesign," Register means also rediscover, re-form, and rewild the city. "Cities existed for 4500 years without cars," he says, and many ancient cities retain old downtowns in which cars are now restricted. Europe has also kept most of its passenger rail system. With this and a generally more conservative view of energy use, Europeans use on average half the fossil fuel as Americans. Register and others in the ecocity movement advocate reestablishing cohesive, walkable, vibrant neighborhoods. They use "opportunity sites" such as abandoned industrial lots and floodplains to create parks, connect services and residences with paths, and open up streams that had long ago been buried in pipes beneath

Bern, Switzerland, one of the European cities that can point the way to a lower-carbon, more humane future. Narrower streets, higher energy prices, dense public transit networks, and long-term commitment to urban living in Europe have encouraged cities such as Bern to move far ahead of most U.S. cities in developing sound energy policies. (SEPTEMBER 2001)

concrete. Vegetated balconies and roofs can take this concept up into the skyscraper realm. Along with a convenient public transport network, the result is that traffic is reduced, pollution lessened, CO_2 actively absorbed, and life made better for residents.

Ecocity concepts are just a few of the ideas being carried out by cities all over the world in response to growing energy costs, high populations, and the specter of global warming. One of the most important changes is to make buses, trains, and other public transport more useful to citizens. Traditional mass transit emits only half the carbon dioxide and nitrogen oxides per passenger mile that cars do. If

New trolley line in the Pearl District, Portland, Oregon. This once-fading area, formerly a warehouse district, has been revitalized with inspiration from European cities and green building techniques. Environment-friendly features of the neighborhood include attractive public transportation, parks, and buildings that save energy with green roofs, solar panels, and highly efficient insulation. Portland has reduced its emissions almost to 1990 levels. (JUNE 2005)

urban Americans used public transit for just 10 percent of their daily travel, says one transit advocacy group, dependence on imported oil could be reduced by more than 40 percent and CO_2 emissions by more than a quarter of the Kyoto guidelines. Nations like the United States that have abandoned much of their rail network need to reinvest in this infrastructure, to link cities and invigorate urban connectivity. Per person or piece of freight, rail systems use only one-eighth the energy expended by cars or trucks. Public transport, which should lead to fewer cars and lower emissions, deserves an equal share or more of subsidies. This includes experimenting with alternative fuels as well. Here, we can follow

Meadow on the roof of City Hall in Chicago, Illinois, one of the first such roofs to be installed in the city. Chicago is now a leader in solar roofs and other green building techniques. This 20,000-square-foot garden, which includes 100 types of native plants, reduces heat radiation from the roof by 5 to 6 degrees and insulates the building. (SEPTEMBER 2002)

the example of nine European cities that are already running buses on fuel cells; three of them, Hamburg, Stockholm, and Amsterdam, make hydrogen with renewable energy.[56]

The world's mass transit champion may well be Curitiba, Brazil—which solved its transportation problem with lots of buses. In the 1970s, facing a growing need for better transportation in this city of just under two million, then-mayor Jaime Lerner had little money and no subsidies to build rail. So the city created dedicated bus lanes, with stops every third of a mile. The buses, which run every minute or so, can transport the same number of people as a subway but are a hundred times less expensive per kilometer. According to Lerner, it is fast and fun to travel in Curitiba.[57]

From the hundred thousand rooftop solar panels subsidized by Shangai to the ten-story Melbourne office building that has PVs, wind turbines, and recycled water cooling or London's Swiss Re office tower (nicknamed the Gherkin) that uses half the energy of a conventional building—around the world programs are at work to cut GHGs and save money. The U.S. Green Building Council's Leadership in Energy and Environmental Design (LEED) program establishes a common standard for integrated, environmentally responsible design, from recycled framing to bike racks and native landscaping. More than six hundred eco-structures, including single-family residences and New York skyscrapers, have been certified by this program (though some participants say it would certify buildings faster if it were less bureaucratic and did not add so much to the cost of construction). Innovative technologies are being written into building codes so that new homes and commercial buildings as well as remodels start out highly efficient. While twenty-two states have some LEED requirements for new public buildings, only a few cities, such as Pasadena, Washington DC, and Boston, have adopted the standards for private construction. Another program, Building America, sponsored by the U.S. Department of Energy, has built or remodeled 33,000 houses that use 30 to 90 percent less energy than standard construction. The goal is to make net-zero-energy homes. This is crucial, since homes and buildings use 40 percent of U.S. energy, much of it for lighting, heating (water and air), and cooling. In Europe one can already buy net-zero pre-fab houses that generate their own power.[58]

With most national governments signed on to the Kyoto Protocol and working toward that 2012 deadline, thousands of localities are working on saving energy and cutting emissions. The Cities for Climate Protection campaign includes 650 local governments worldwide. Many of the most successful urban conservation and energy use techniques come from Europe, where towns were built for small vehicles and pedestrian traffic and have survived energy shortages many times. In a drive started by Seattle mayor Greg Nickels, at least 900 American mayors representing millions of residents in fifty states have pledged to reduce greenhouse gas emissions in their cities to 7 percent below 1990 levels by 2012. Meanwhile, just down the coast from Seattle, Portland, Oregon, has already gone far along that path. It was the first U.S. city to have a climate action plan, and since 1993, using conservation, energy management, building code efficiency standards, and transportation planning to reduce car travel, it has reduced its greenhouse emissions to near 1990 levels. Car traffic over Portland's bridges is slightly less than in 1991, while two and a half times more bikes are crossing. Urban energy and emission savings have also come from using low-wattage lighting and traffic signals, alternative-fuel vehicles, and light rail lines. On this continent, Chicago and Toronto lead in green roofs, an innovation that is al-

GLOBAL WARMING ON TRIAL

Courts are increasingly being called on to judge the interaction between governments and people on the issue of climate change. As global warming becomes more widely understood, getting court rulings may be a faster route to change than negotiating laws and international agreements.

In the most momentous global warming case in the United States, *Commonwealth of Massachusetts v. EPA,* the Supreme Court ruled that the Environmental Protection Agency has authority under the Clean Air Act to regulate carbon dioxide from vehicles as an air pollutant. In a 5 to 4 decision in April 2007, the Court held that the agency must provide reasoned and scientific justification for refusing to regulate. Although it could not order the EPA to act on CO_2, the Court's majority opinion was that there is little scientific doubt that vehicle emissions are contributing to the "serious and well recognized" harms from climate change.

Massachusetts, eleven other states, and a roster of other jurisdictions and environmental groups sought the ruling after the EPA refused to regulate tailpipe CO_2, asserting that the science on global warming was too uncertain. The Supreme Court allowed Massachusetts to sue based on harm to its land from conditions such as rising sea level, which it said was made worse by federal inaction. The judgment in *Commonwealth of Massachusetts v. EPA* will now become precedent in other CO_2 regulation cases whose outcomes turn on the federal role in climate regulation. In *State of New York v. EPA,* for example, states and environmental groups are seeking to force the agency to control carbon dioxide from power plants.

Another case that has been affected by the Supreme Court decision is *Central Valley Chrysler-Jeep v. California Air Resources Board,* a suit brought by automakers against a California law requiring cuts in CO_2 emissions from motor vehicles. The regulations give car companies until 2016 to achieve a 30 percent reduction in greenhouse gas emissions from new cars, pickups, minivans, and SUVs. The automakers claim that California may not curb vehicle emissions of greenhouse gases because it is actually setting mileage standards—an action that is reserved for the National Highway Traffic Safety Administration. California says it is regulating air pollution, which it is allowed to do by EPA waivers. Automakers already have lost a similar case over a Vermont law modeled on California's. California has filed a motion that the Supreme Court decision resolves the case in the state's favor.

California has duly applied for a waiver from the EPA for this law, but in December 2007 the EPA Administrator, Stephen Johnson, denied the permit, giving no reason. California is now suing the EPA over this refusal (*California v. EPA*).

Meanwhile, Johnson, rather than set rules controlling CO_2 under the Clean Air Act in accordance with the Supreme Court ruling, bowed to apparent intense White House pressure not to act. Instead he issued an "advance notice of proposed rulemaking," which was an elaborate request for public comment, effectively delaying action until after the presidential election.

Other lawsuits are underway to force labeling of power plant CO_2 emissions as public nuisances, to hold automakers liable for monetary penalties for the greenhouse gases emitted by their products, and to challenge U.S. vehicle mileage regulations for SUVs and light trucks, which some see as an incentive to continue building giant gas guzzlers.

Legal petitions are being issued on the international level as well. The Inuit Circumpolar Conference, an international organization representing approximately 150,000 Natives in the Arctic regions of Alaska, Canada, Greenland, and Russia, filed a petition in 2005 with the Inter-American Commission on Human Rights. The petition charged that the United States is violating Natives' rights to the benefits of culture, property, health, security, and subsistence by increasing global warming. After the petition was denied in November 2006, the Inuit Conference took the issue to the United Nations, which passed a human rights resolution in March 2008.

Sources: I thank attorney Frances Raskin for research and background information. See *Massachusetts v. EPA,* April 2, 2007; Linda Greenhouse, "Justices Say E.P.A. Has Power to Act on Harmful Gases," *New York Times,* 3 Apr. 2007; Felicity Barringer, "2 Decisions Signal End of Bush Clean-Air Steps," *New York Times,* 12 Jul. 2008; "Advance Notice of Proposed Rulemaking: Regulating Greenhouse Gas Emissions under the Clean Air Act," at www.epa.gov/epahome/pdf/anpr20080711.pdf; Amanda Leigh Haag, "Going to Court over Climate Change," *Nature News* (8 Sept. 2006), at www.nature.com/news/3006/060904/ full/060904-16.html; For the Inuit petition, see www.earthjustice.org/our_work/cases/2005/inuit_human_rights_and_climate_change.html; also Richard Sieg, "At International Commission, Inuit Want to See Change in U.S. Policy on Global Warming," *Vermont Journal of Environmental Law* (2 Mar. 2007), at www.vjel.org/news/NEWS100058.html; Stephen J. Porter, Center for International Environmental Law, email 21 Jul. 2008. ✦

ready quite common in European cities. Salt Lake City is capturing methane at the landfill and at water treatment plants.[59]

IMPLEMENTATION WITHOUT RATIFICATION

American leaders at the state level, too, are beginning to take note of the science of global warming, see the effects in their regions, and proceed toward emissions reduction policy and legislation. Because governors and state administrators can influence broad sections of the U.S. population, these actions are sometimes called "implementation without ratification"—referring to the refusal thus far of the federal government to take part in the Kyoto Protocol or require greenhouse gas reductions. More and more leaders, however, believe that what is really needed is a national policy, because uncoordinated state-by-state actions might "increase the uncertainty facing the business community, thus potentially making the most cost-effective solutions more difficult." As a new president and Democratic Congress take up federal action, more than half the states, led by New England and California, have programs in place to reduce greenhouse gas emissions, and twenty-three plus the District of Columbia have requirements for utilities to generate some part of their electricity from renewable sources. Even Alaska, long a holdout because of its dependence on oil revenues and staunchly skeptical leadership, has a new law requiring study of climate effects and solutions. The Regional Greenhouse Gas Initiative, launched by former New York governor George E. Pataki, links nine states from Maine to Maryland in developing a regional cap-and-trade program to limit carbon dioxide emissions from power plants.[60]

The strongest force in state climate change initiatives is California. The state has an economy larger than all but six nations and is the world's twelfth largest source of CO_2. In August 2006 it set itself apart from the rest of the nation and most of the world by passing a law to limit greenhouse gas emissions across the board to 1990 levels by 2020. The law was passed by a state legislature bolstered by predictions from sixty economists, including three Nobel Laureates, that capping emissions was "potent strategy" for driving not only reductions in global warming but also job growth. One prediction was that the efforts and invention encouraged in all segments of the state will boost the economy by more than $60 million and create between 17,000 and 89,000 jobs in the coming years. The limits, to begin going into effect in 2012, will mean a 25 percent reduction in greenhouse gases by 2020, dropping further to 20 percent of 1990 levels by 2050. The Democrat-led legislature wrote into the bill protections for low-income communities and a system to ensure that CO_2 cuts are permanent, quantifiable, and verifiable. The new law pushes ahead by thirty years reductions that Governor Arnold Schwarzenegger had instituted in 2005. At that time, the governor declared, "I say the debate is over. We know the science, we see the threat, and the time for action is now." With most state greenhouse gases coming from tailpipes, in 2002 the Golden State tightened automobile pollution rules, giving car companies until 2016 to achieve a 30 percent reduction in emissions from new cars, pickups, minivans, and SUVs sold in the state. The American car industry immediately sued, but ten other states followed California with similar exhaust requirements.[61]

The California legislation will certainly not be the last to require manufacturers or energy providers to cut the emissions of their products. Some level of federal greenhouse gas regulation now appears certain. But with so much at stake, corporations should not wait for laws to be passed to force their hand. Sixty percent of U.S. energy is wasted between its sources and end use. Companies of all sizes thus have a great opportunity to stem this waste—moreover, at a profit. Physicist Amory Lovins, a major advocate of increased energy efficiency, writes, "If properly done, climate protection would actually reduce costs, not raise them. Using energy more efficiently offers an economic bonanza—not because of the benefits of stopping global warming but because saving fossil fuel is a lot cheaper than buying it." In a virtual library of papers and reports over the past thirty years, Lovins and colleagues at the Rocky Mountain Institute have reenvisioned American industrial energy use at a much higher level of efficiency.[62]

According to their studies, in the average car only 13 percent of the fuel energy reaches the wheels, and much of that is lost to flexing and heating of the tires. The same inefficiency marks industrial processes. Losses between a power plant and factory output—in transmission lines, motors, pumps, pipes, and so on—can waste more than 90 percent of the energy input, and as much as 70 percent is lost at the power plant itself. In some cases, just using bigger pipes that run in straighter lines can make for large savings. The Lovins team champions eliminating the risks and inefficiencies of giant central power plants in favor of decentralized power, near the point of use. "This isn't rocket science," Lovins wrote in 2005. "In all, preventable energy waste costs Americans hundreds of billions of dollars and the global economy more than $1 trillion a year, destabilizing the climate while producing no value." He estimates that "adoption of efficient vehicles, buildings and industries could shrink projected U.S. oil use in 2025 . . . by more than half." This may be optimistic, especially since his idea for hypercar lightweight composite vehicles appears to be so distant in Detroit's rearview mirrors. The Rocky Mountain Institute also recommends other steps to cut petroleum use, including using creative business models and policy to speed the adoption of new transportation technology, replacing a quarter of oil with biofuel, and conserving and redirecting natural gas, perhaps converting it to hydrogen. Lovins sees the carrot for this shift (and for the adoption of alternative fuels generally) in stark economics: "Saving each barrel of oil through efficiency improvements costs only $12, less than one-fifth [one-eighth, as of late 2008] what petroleum sells for today."[63]

One company that has made an effort is British Petroleum, BP, which between 1997 and 2001 cut its emissions by 10 percent, spending $20 million to do so but netting $650 million in savings. Chairman John Browne, who used a sort of internal "cap-and-trade" market to achieve the cuts, advocated the idea of the green corporation confronting global warming. BP—the ads say it now stands for "Beyond Petroleum"—claims it is committing hundreds of millions of dollars to researching alternative fuels and energy. Critics and some environmental nongovernmental organizations could be forgiven for questioning its motives when the preponderance of business has remained in oil and it was responsible for the largest-ever oil leak on Alaska's North Slope in 2005 and a serious pipeline rupture in 2006. Its website declares: "Our goal is to build a profitable, global and market-leading low-carbon power business by 2015. By this date, we estimate that this will help to reduce forecast greenhouse gas (GHG) emissions by 24 million tonnes a year." Although this is less than 2 percent of its overall emissions (direct and from product use) of more than 1.3 billion tons per year, changes must begin somewhere, and

BP has done more than most oil companies. A notorious counterexample is ExxonMobil, which has heavily funded political, "scientific," and media spokespeople to downplay global warming and sow confusion about the strength of the underlying science.[64]

In December 2000, an annual survey was launched that helps companies worldwide assess their role in climate emissions. Called the Carbon Disclosure Project (CDP), it is currently sponsored by 385 institutional investors controlling some $57 trillion in assets. In 2007 it contacted 2,400 of the largest corporations in the world by market capitalization, asking them to take steps to reduce their emissions and report exactly how and by how much. Companies are asked to disclose investment-relevant information concerning their pollution, including efforts to curb emissions, to address physical risks connected to climate change, and to develop new, "green" business opportunities. In 2006, 72 percent of the companies on the CDP's list replied to the survey.[65]

Ceres is a coalition of financial, civic, and environmental groups helping businesses move toward cleaner operations and sustainability. Their Global Reporting Initiative (GRI) is used by over 1,200 companies for corporate reporting on environmental, social, and economic performance. Ceres encourages businesses to manage and plan for climate risk and opportunity. Publications point out to CEOs and board members the likely financial impacts—physical, regulatory, competitive, and reputational. Every business plan, Ceres counsels, must include openly disclosed greenhouse gas inventories and reduction targets, with managers looking far upstream and downstream to see the climate "footprint" of their company and research solutions.

Chevron, which also replied extensively to the CDP survey, is the first U.S. oil company to disclose its entire footprint, including how its products are used. According to Ceres, Chevron's efforts began with the development of a greenhouse gas emissions inventory method that established how and what to report. The inventory requires metering or measuring gas flares, methane releases, and thermal values. Chevron also commissioned a third-party verification of its inventory. In 2004, using the inventory as a planning tool, the company reported reducing emissions by 1.4 million tons of CO_2. Among the corporations with the most complete climate disclosures and programs are HSBC Bank, Unilever, RWE Utility, BP, and Rio Tinto. In addition to taking this voluntary action, some companies, among them Wal-Mart, Shell, General Electric, and Duke Energy, along with several other huge energy firms, told a Senate committee in 2006 that they expected and would accept mandatory caps on the amount of carbon they sent up. Along with Johnson & Johnson, Nike, and IBM, they joined with citizen groups to propose voluntary and legislative action.[66]

HIGHER EFFICIENCY, LOWER EMISSIONS, INCREASED PROFIT

Cap-and-trade is familiar to business from the U.S. sulfur dioxide acid rain regulations: the amount of pollution is limited by law or legal agreement, and those firms that cut their output soonest can make money selling their unused allowance to companies that are slower to meet their goals. The cap on total emissions decreases over time, accomplishing the needed reduction. When CO_2 reductions are valued or regulated, carbon markets are an obvious mechanism for meeting those goals. The Chicago

Climate Exchange, established in 2003, was the first voluntary greenhouse gas trading system, in which members are committed to overall reductions of 4 percent below 2001 levels by 2006 (now extended to an additional 2 percent by 2010). Among its more than 350 members are cities, colleges, and farm bureaus, as well as such corporations as Ford, DuPont, International Paper, and IBM. The Emissions Trading Scheme, set up by the European Union under the Kyoto Protocol and incorporating about 15,000 industrial sites, saw growth to beyond $50 billion in 2007. It is by far the largest carbon allowances market. Prices fell a year after Kyoto limits went into effect, however, when some members' actual emissions for 2005 came in less than they had told the market to expect. There was thus a surplus of emissions credits—the cap was too "loose," making it less necessary for heavy polluters to buy extra polluting allowances. Some claimed this was an effort by some countries to allow their industries higher limits, but other analysts saw just the birthing difficulties of an international carbon market.[67]

Meanwhile, in another procedure for reducing emissions based in the Kyoto Protocol, industrialized nations will pay for new projects that reduce emissions in developing nations. The rich countries then get credits against their mandatory Kyoto goals. Part of the idea was that poor nations would be helped by this Clean Development Mechanism. So far, most of the 1,000 projects are in just four countries: India, Brazil, Mexico, and China. Almost half the total allowance payments went to China for reducing a very potent by-product of an ozone-depleting hydrochlorofluorocarbon refrigerant. The waste gas, HFC-23, is 11,700 times as strong a greenhouse gas as CO_2, so it makes a good target for reduction. However, its quantity is not great compared to that of CO_2, CH_4, and NO_2, which make up 88 percent of the greenhouse effect. Analysts and environmental groups point out that the prices paid to destroy HFC-23 under Kyoto rules are so high that the two Chinese companies involved will make a large profit, along with all the financial brokers of the deal. This, according to one nongovernmental organization, creates an incentive to make more of the refrigerant, which the Montreal Protocol is trying to phase out. The idea of the two protocols working at cross-purposes while so much money flows to China and very little to poor nations was a topic of concern at the 2006 climate talks in Nairobi. African leaders encouraged broader participation, money for adapting to changes, and clean development projects such as nonfossil power for their continent.[68]

The Kyoto mechanism is set up to encourage cooperation between rich and poorer nations to make life better for people around the world by actively sharing technology. The common practice of passing to developing nations the previous generation of industrial equipment, no matter how polluting or inefficient, locks into place gigantic amounts of greenhouse emissions. Technology transfer allows less developed nations to leapfrog the old ways and use new inventions and techniques for reducing carbon output. The Kyoto Protocol includes mechanisms to assist in this transfer, including Joint Implementation, whereby developed nations work together to bring technology to other countries.

Another way that money is flowing is in the direct funding of energy and efficiency innovation by venture capitalists. It is clear to a growing number of financiers that the drive to help slow global warming will generate large numbers of inventions and new businesses—much as the computer industry did. Worldwide, renewable energy is a $148 billion business. As one of the prime movers in this field puts it, "Policy and innovation will solve these problems, with leadership from entrepreneurs who want to build profitable ventures and do good at the same time." This company, Kleiner, Perkins, Caufield & Byers, is credited with helping convince California governor Schwarzenegger to sign the new emissions-limiting legislation. In 2005 the Goldman Sachs Group, one of the leading investment banks, pledged to become a market maker in climate-related commodities as well as a leading U.S. wind energy developer and generator, and to make up to $1 billion available for renewable and energy efficiency investments. As the *Economist* and others have warned, however, renewable enterprises could fizzle the way many dot-coms did. New companies and new power sources are going to have to withstand a very unstable time of changing laws, shifting prices, competition from oil and gas, and unequal subsidies.[69]

Within this setting, a significant number of businesses, among them many energy and transportation companies, insist that the science behind global warming is still too uncertain, or else that the regulatory framework is too inconsistent to guide their actions. Other companies appear to have advanced advertising far more than actual change. Strong businesses don't wait long in seeking to control their production and marketing environments; they change and innovate. All business leaders must see how much global warming affects and is affected by them—and seize the initiative.

ONE MAJOR SECTOR of the world economy that is particularly troublesome is air travel and freight, which is growing rapidly and probably cannot be powered by noncarbon fuels, at least not in the near future (though development of bio-kerosene as a jet fuel is being discussed). Jet engines run on a type of kerosene, which is similar to gasoline in the amount of heat-trapping gases it produces when burned. The more than 24 million flights per year contribute 1.6 percent of world CO_2, 5 percent of the water vapor, plus measurable amounts of nitrogen oxides (which can form ozone) and sulfur compounds— and they spew much of this in the upper troposphere and lower stratosphere. At these heights, 25,000 to 40,000 feet (7600–12,200 m), the greenhouse gases have about three times the effect they would have at ground level. Ozone is especially potent at this altitude. Jets also commonly create condensation trails (contrails) at high altitudes, many of which are persistent and spread into a type of cirrus cloud. Although these thin high clouds may cool the Earth a little during the day, at night they hold in heat, becoming a measurable addition to global warming.[70]

Although per person, airplanes emit slightly less CO_2 on long flights than would be created if those people drove the same distance alone, with two or more people in a car, driving is less polluting than flying. Short flights of 300 miles (500 km) are twice as polluting per passenger. According to the International Civil Aviation Organization, scheduled commercial flights covered more than 17 billion miles (23.4 billion km) in 2004, carrying close to two billion passengers and 37.6 million tons (34 million mt) of air freight. Most of that flying is not inherently urgent; it's just what we got used to in a few years of cheap energy. Some of this transport could be accomplished by less polluting surface modes, at least within continental borders. Business travel could be reduced by greater use of telecommunication. Sixty percent of air travel is for tourism, which is not strictly essential, despite its promise of world understanding and, in some places, jobs. In the short term, giant planes like the Airbus A380 can carry more people per flight, while Boeing's new 787 series promises greater fuel economy. Many of these measures

This field on an industrial farm near Davis, California, is ready for planting. The farming practices used here are typical of those used by corporate agriculture in the United States and in some other nations. These practices are fossil fuel intensive, not only because of the machines needed to maintain monoculture, but also because most fertilizers and pesticides are made from petroleum and natural gas, and because the food products tend to be shipped long distances to market. (JUNE 2003)

will be forced as the European Union brings the aviation industry into its mandatory emissions cut scheme. Transport by ship is not a viable alternative to air freight, as studies have found that the 90,000 merchant freighters operating worldwide burn a high-sulfur bunker fuel that may cause 5 percent of world carbon emissions.[71]

CLIMATE CHANGE BEGINS AT HOME

Cutting across all the parts of our world that use energy and fuel, as well as all the technological fixes and changes that need to be investigated and instituted, is the simplest, cheapest, and most effective

Full Belly Farm, a large organic farm north of Davis. Organic and permaculture farming practices use far less energy, preserve the carbon storage capacity of soils, and promote better health than conventional farming. Consumers can encourage organic farming—and eat better—by buying at farmers' markets and at stores offering certified organics and local produce. (JUNE 2003)

way to cut greenhouse gases: efficiency. Old-fashioned conservation. No energy alternative is totally innocent of CO_2, except *not* using electricity or fuel. David Pimentel and colleagues, in a 2002 update of an energy survey originally undertaken in 1994, wrote that the first priority "should be for individuals, communities, and industries to conserve fossil fuel resources by using renewable resources and by reducing consumption." "Efficiency improvements," says Harvard's John Holdren, "are the cheapest, cleanest, surest, most rapidly expandable energy option we have." Everyone can do it, and everyone must.[72]

With help from environmental groups and utilities, home owners and motorists in the United States and other industrialized nations have begun to take initial steps to save energy, such as installing better windows and insulation, using efficient lightbulbs and appliances, buying higher-mileage cars, driving less, or simply driving a little more slowly. These ideas have been around for many years as ways of saving money, but now they have much more importance for reducing global warming. Worldwide, efficiency actions could reduce the annual increase in energy use to under one percent. Operating a typical American house can produce 26,000 pounds of greenhouse gases a year, according to the Rocky Mountain Institute. Switching to compact fluorescent (CF) lights has become the cliché of personal action against global warming. However, a single bulb changed to a CF saves some $50 off electric bills over the bulb's five- to six-year lifetime. Replacing most of the bulbs (say, an average of forty) in most American homes could eliminate 260 million tons of CO_2—more than the combined emissions from America's twelve largest coal power plants. The 2007 Energy Bill encourages this switch. The standby mode on entertainment equipment and various transformers for low-voltage lights and cell phone chargers cost the United States 6 percent of its electrical energy. Instead, we can plug these into power strips with on-off switches or into a switched circuit. Cutting back on hot water use through low-flow shower heads and using dishwashers and clothes washers at lower temperatures and only when full are also great savings. Cleaning the refrigerator's condenser coil and setting the dial even a few degrees warmer can cut its electric use by up to a third.[73]

The food in that refrigerator is much more than it seems, for when we devour an apple or a chicken breast, we are also devouring energy. As we know from the above discussion of the quest for nonfossil fuels, agriculture runs on oil. By one estimate, 17 percent of all fossil fuel used in the United States goes for food growing, harvesting, processing, shipping, packaging, and distribution. The average bite of food has traveled about 1,500 miles (2400 km) from the field to your table. Thus, eating locally grown and organic food can save energy and emissions as well as improve nutrition. Wal-Mart's decision to carry organic foods (if the scale of chemical-free agriculture it demands does not dilute organic standards) could reduce dependence on fossil fuels. One study found that organic corn production took 30 percent less fossil energy than conventional farming. Organic and water-saving planting techniques can also restore carbon to the soil and help keep it there. Even more shifts are due in the livestock industry, including grazing and feed improvements, methane and waste control, alternatives for overgrazed areas, and consumer diet changes. Overcoming farming's reliance on greenhouse gas–intensive techniques, oil-burning machines, and chemicals will require international planning and cooperation, but it must begin near home.[74]

IN THE BIG PICTURE our energy predicament is also due to grand assumptions of international economic policy that value consumption, constant monetary growth, and cost-benefit analyses favoring the present over the future. Economic decisions today are often biased because environmental and social costs are treated as of secondary importance; they are either not counted or pushed off on future generations and other parts of the world. In the fall of 2006 the *Stern Review*, a 700-page British economic report, directly confronted economic assumptions about global warming. Headed by the former chief economist of the World Bank, Sir Nicholas Stern, the report is a serious wake-up call to the world's leading nations and their financial leaders. "The benefits of strong, early action on climate change outweigh the costs," Stern says. "An urgent global response" is required. "Tackling climate change is the pro-growth strategy for the longer term," the report states. Stern calls the traditional economic models that have guided nations and corporations "the greatest and widest-ranging market failure ever seen": there is little or no cost to spew out carbon, but its effects will devastate economies. Stern advises taking substantial steps to limit emissions and climate change, at a cost he estimates to be about $600 billion per year (in today's dollars), or one percent of the total world GDP.[75]

This cost is an investment "to avoid the risks of very severe consequences in the future." "The evidence shows that ignoring climate change will eventually damage economic growth," the report says, and cause disruption "on a scale similar to the great wars and the economic depression of the first half of the twentieth century." Taking into account estimates of runaway climate change, the report predicts a shrinkage of GDPs and individual consumption levels of between 5 and 20 percent for several centuries, if little action is taken.

There have been many critiques of the report, of course, some focused on this last computation, which is apparently based on the highest levels of possible effect if the world does nothing at all about global warming during this century. The world is already starting to react, and many scientific estimates put temperature rise and effects in the center of the range of possibilities. But the damages are still likely to be hugely expensive. Among those who offered scathing comments was Richard Tol of Hamburg University, who according to the *Economist* called the report "alarmist and incompetent." One of the most respected economists, William Nordhaus of Yale University, published a forecast earlier in 2006 for a 3 percent loss of economic output from a doubling of CO_2 and its projected effects. Nordhaus's major criticism of the Stern numbers appears to be in the social discount rate—a somewhat arcane calculation having to do with how much we value the people in the future who will have to deal with the drastic changes, compared to the people alive today. Stern values both groups equally, so he would make very drastic cuts in fossil fuel emission now to try to save the future generations. Nordhaus and others believe that if economic growth continues, there will be more technology in the midterm to help lower emissions, so that we today do not have to take such drastic measures all at once.[76]

The *Stern Review*'s central meaning, however, is that we should act now precisely because there is a possibility of disaster, and that it would be highly unethical to ignore it. Within the report is a rather rare thing in economics: a long discourse on ethics. "Many would argue that future generations have the right to enjoy a world whose climate has not been transformed in a way that makes human life much more difficult; or that current generations across the world have the right to be protected from environmental damage inflicted by the consumption and production patterns of others." This statement conforms nicely with words from the climate change treaty that nations "should protect the climate system for the benefit of present and future generations of humankind, on the basis of equity."

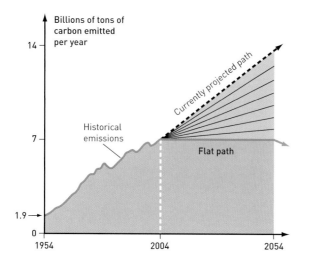

Billions of tons of carbon emitted per year

14 —

Currently projected path

7 —

Historical emissions

Flat path

1.9 →

0 —

1954   2004   2054

WEDGES

- Increase efficiency of vehicles
- Reduce use of vehicles
- Increase efficiency of buildings
- Increase efficiency of coal plants
- Replace coal power with gas power
- Capture and store CO_2 at coal plants
- Replace coal power with nuclear power
- Replace coal power with wind power
- Replace coal power with solar power
- Capture and store CO_2 at hydrogen plants
- Capture and store CO_2 at coal-to-synfuels plants
- Use wind to generate hydrogen for fuel-cell cars
- Replace fossil fuel with biomass fuel
- Reduce deforestation
- Expand conservation tillage

Carbon emissions can be stabilized, and eventually reduced, if we make changes now. The "stabilization triangle," conceived by scientists Robert Socolow and Stephen Pacala, shows how reductions could be realized over the next 50 years to keep emissions at present-day levels. Each of the technologies listed on the right represents one "wedge" capable of reducing carbon emissions by 25 billion tons over 50 years. Seven such wedges would keep emissions at current levels. More wedges, including others not listed, can be implemented to reduce emissions even further. (ADAPTED FROM ROBERT SOCOLOW AND STEPHEN PACALA, "THE URGENCY OF CARBON MITIGATION," 2006)

EDGES AND WEDGES

The world we live in—with all its promise and failure, with the millions who enjoy the benefits of energy use and the billions who do not—is at the edge of a precipice. Oil supplies are beginning to be stressed, and although a long transition back to coal as the major energy source might seem to be a solution, we see that the cliff edge is sharp and the chasm deep. The crisis of global warming caused by our use of fossil fuels will escalate rapidly if we maintain even our present use. Rather than continuing in the direction of using oil and gas, we must step back from the edge and take another path. The analogy that many commentators are beginning to use as shorthand for our situation is attributed to Sheikh Zaki Yamani, former Saudi Arabian oil minister: "The Stone Age came to an end, not because we had a lack of stones, and the Oil Age will come to an end not because we have a lack of oil."

Sheikh Yamani reached that conclusion in 2000, as quoted by Amory Lovins, because he saw that inventions like fuel cells "will cut gasoline consumption by almost 100 per cent" and that "on the demand side there are so many new technologies." The wide variety of actions that lower carbon emissions and increase efficiency are encouraging, but as the authors of a survey of them in *Science* said, "Combating global warming by radical restructuring of the global energy system could be the technology challenge of the century. . . . At the very least, it requires political will, targeted research and development, and international cooperation."[77]

As Lovins wrote, "energy policy harms the economy and the climate by rejecting free-market principles and playing favorites with technologies." U.S. government greenhouse gas research and reduction programs are already in place that could easily be strengthened and expanded. Redirecting fossil fuel tax breaks and research funds toward renewables would pay great dividends, as would eliminating perverse subsidies and laws, such as those that lower taxes for buyers of SUVs or allow coal power plants to discharge mercury into the air. Ross Gelbspan has suggested a financial transaction tax that could raise $300 billion a year for climate protection projects. Al Gore advocates taxing greenhouse emissions rather than income. Instituting a mandatory cap-and-trade plan for industrial greenhouse gas reductions would be a great boon, as would providing incentives for reducing fossil fuel use. Product labels with carbon amounts involved in manufacture and use, and residential "smart meters" showing real-time energy cost, would keep the issue on everyone's agenda. These proposals don't require technological breakthroughs. They require leaders and legislative bodies to just do it.[78]

With policy solutions and technological innovations, we've been dealt a winning hand; each card makes a significant contribution to a whole solution, but none can or need be sufficient unto itself. Consider a plan by Stephen Pacala and Robert Socolow of Princeton University that posits "stabilization wedges," fifteen existing technologies, each of which can prevent 25 billion tons of carbon output over fifty years. The conservation and efficiency wedges are already very doable; others, such as carbon sequestration, will be less certain. Under one scenario, only seven wedges could halt CO_2 rise at around 500 ppm by midcentury. But the key is to start soon. Bringing U.S. emissions down from 24 to 5 percent of the world total by midcentury could happen with continuous cuts of no more than 3 percent a year if we started now, says the Natural Resources Defense Council. Waiting twenty years to start would then require 8 percent annual reductions. A plan written for the U.S. Energy Department in 2000 called for efficiency, replacement of polluting power plants, more research and development, and tax incentives in order to cut energy use by 20 percent and greenhouse gas emissions by 30 percent in twenty years. There are important plans published by the National Commission on Energy Policy in 2004, the Pew Center on Climate Change "Agenda for Climate Action" in 2006, and many others.[79]

To investigate, understand, and invent what needs to be done, much more money must be spent on climate, mitigation, and energy research. The great pulse of American research in response to the energy crisis of the 1970s peaked near 1980. With Ronald Reagan's presidency, spending on energy research and development plummeted, and it has stayed at a low level ever since (currently about $4 billion a year). Many other national budgets similarly shortchange this area. It has been suggested that seven to ten times more be spent and that an organization akin in size and level of support to the Manhattan Project (developer of the atom bomb in the 1940s) or ARPA/DARPA (the [Defense] Advanced Research Projects Agency, originator of the internet between 1969 and 1972) be established to focus on the energy/global warming problem. Funding should also be restored, at increased levels, for NASA's Earth-observing program, which is currently suffering cuts (partly in favor of the Space Station and a new moon mission). To help feed the innovative spirit, communication among scientists, government leaders, and the public must be unfettered. Scientific findings need to be integrated with clear and detailed forecasts of what may happen and how we plan to deal with it. Support for basic science education from grade school up will help everyone to understand the situation and feed the universities with scientists for the future.[80]

NEROS OR HEROES?

Are all these actions going to be easy to implement? Most of them, no. Can we anticipate difficulties in everything from design to training to public acceptance? Of course. Will improvements occur uniformly and benefit everyone at once? Not a chance. Corporations face transition costs and risk sacrificing a competitive advantage in an already tough world market. This effort will take dedicated and bold leadership, nationally and worldwide, across a broad political spectrum. But the benefits are immense, and the goals are worthy in their own right. As John F. Kennedy said in 1962, advancing his space program: "We choose to go to the moon in this decade and do the other things, not because they are easy, but because they are hard, because that goal will serve to organize and measure the best of our energies and skills, because that challenge is one that we are willing to accept, one we are unwilling to postpone, and one which we intend to win, and the others, too."[81]

Considering the threat that global warming poses to our world and to human civilization, the response we make is much more important than was the Apollo moon mission and even the Marshall Plan of European reconstruction after World War II. Some of these great campaigns are linked because they involved recognizable strong threats or goals and powerful leadership that could tip the balance to action. It is perhaps obvious, but needs to be recalled nonetheless, that the leaders who called us to action in the past did not themselves know how to build warships, make moon rockets, or craft technical legislation. There were doubts and disagreements galore, as well as plenty of incomplete information. But leaders were able to find men and women who could do what needed to be done and rally the rest of us to support them in a vision of progress born of threat, whether that threat was to the nation, to social equality, or to economic viability. We have threats today, too; we have much progress to make on human rights and health. But instead of fight-to-the-death international competition we have a world largely ready to cooperate on climate change.

As for what to do about it, we again have had questions and incomplete information. Many factors influence how the atmosphere warms, and how we perceive the results of that warming. Doubts, however, are fading. There is no doubt that putting more CO_2 into the air will raise the atmospheric temperature; this scientific theory has been tested and accepted. How fast temperatures will rise is predictable within limits determined by the composition of atmosphere and ocean. How much warmer it will get in Washington DC or Tuvalu, and what the precise effects will be—that we cannot know for sure. Any individual study of the consequences of climate warming should and will be questioned and tested by other researchers. Climate scientists deal with uncertainties daily and engage in ongoing discussions in scientific journals.

By now, though, there are thousands of repeat, independent, careful observations of effects on ice, plants and animals, mountains and deserts, the air and oceans, and they lead to a sense of certainty that climate change is unequivocal. This is not just about how many scientists agree; there are a few who have strong reservations in certain areas. It is, rather, about the quality of the research and weight of

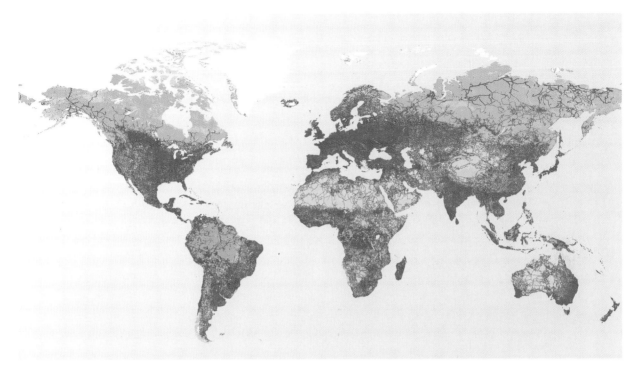

A depiction of human impact on the Earth, 2002. Red areas and lines indicate the locations of settlements and modern infra-structure such as roads, powerlines, and pipelines. (COURTESY GLOBIO/UNEP)

the results. We know more about what is happening with global warming than we do about stocks, business plans, and, apparently, military planning. Historian Spencer Weart has written that fifty years ago, when rapid greenhouse warming was first conjectured, scientists, as "professional skeptics," mostly did not believe it was a big problem. "It took decades of accumulating evidence," he said, "with many hard-fought debates, to convince them they were wrong." Today, as entomologist Camille Parmesan told an interviewer, "The scientists are not having a debate. The debate is among the politicians."[82]

Getting past uncertainty and moving on to action that reflects the values and most pressing needs of the nation is the number-one job of politicians. It requires judgment and precaution. Home and business owners take precautions not to start fires, but they also invest in insurance because they know complete prevention is unattainable. "Politicians have to act in the face of uncertainty," said Nobel Laureate Mario Molina, a comment that was echoed by Klaus Toepher of the UN Environment Programme: "We will never in human situations decide in certainty. We are always deciding with limited resources and information." The world needs more leaders who can understand and articulate the threat of climate disruption, and then translate the message of science into actions that will advance health, progress, and democracy, while safeguarding the precious world we live in.[83]

Absent this leadership, many values, both personal and national, are jeopardized:

Strength and national security, including response to weather and disease emergencies, are weakened by dependence on and debt from foreign oil. Terrorism, isolationism, and economic vulnerability increase, as does the threat of war.

Self-sufficiency diminishes with our growing dependence on oil as a primary fuel and on other nations to provide it. The United States spends about $1 million a minute on imported oil. National coal reserves are vast, but burning coal poisons our climate and our health. Meanwhile, huge resources in solar, wind, and biologic energy, and huge potential savings from greater energy efficiency, remain largely untapped and underfunded.[84]

Good government at all levels is threatened by the consolidation of energy ownership by corporations and by OPEC, as well as by military actions to secure foreign supplies and defend against terror. Political systems may be corrupted and other national and regional potentials may remain unrealized. Communities would be strengthened by decentralized energy sources.

Finally, there are the values of justice, morality, and righteousness: the ethical realm. The effects of climate change are distributed very unfairly, as are the profits and costs of energy extraction and use. Energy costs sap the coffers of individuals and nations as a whole, reducing the money available for other needs, including education. Steadfast religious convictions should guide us to care for the Creation. Honoring God's gifts and treating others as the children of God are central tenets of religion but are often betrayed by priorities driven by economic needs.

Indications are strong that dangerous change is near. Yet because the United States and leading economic powers—who happen also to be the premier polluters—want to ensure no harm to their economic interests, they fail to respond to climate change as the deep crisis it is. This inaction may not only be self-limiting; it will soon seriously harm some of the smallest nations and poorest people of the world. We need to address this powerful ethical challenge and begin to create a more balanced, respectful, cooperative world society.[85]

We are dependent on fossil fuels for our most basic needs—working, putting food on our tables, and housing our families. But fossil fuels also propel the excesses of consumer culture. Fewer than one person in ten in the world owns a car. Yet we in wealthy nations can drive alone in a two-ton mass of steel and plastic a short distance to a mall that is made of fifty acres of concrete, to buy oranges and running shoes and coffee that were flown thousands of miles to be there for us. Cheap petroleum has given us the freedom of mobility, but the cost of this mobility in air pollution, endless suburbs, crowded freeways, and diminished wilderness is high. Our materialistic way of living is being emulated around the world. The science of global change, the economics of energy, and the finite reality of the resources of the Earth say it can't continue. What should be the vision for our future? A better direction would emphasize stronger communities, better home life, cleaner air and water, less-expensive utilities, expanded local businesses, world health, human rights, and international cooperation, and a planet that can support all of us. Many see working together to limit global warming as a major step toward peaceful collaboration among the world's nations. "What we are discussing in discussing climate change is the peace program of the future," said Klaus Toepher of UNEP, "the disarmament program of the future."[86]

Despite the difficulty of creating a new energy future, who would not want to move in this direction, with all its benefits? The greatest responsibility citizens have is to vote, with their money and their ballots, for changes in the products and policies offered to them. Imagine if no one bought a vehicle until they could have what they needed in a 70-mile-per-gallon plug-in hybrid running on easily available cellulosic ethanol. Imagine if cars were not really needed much anyway.

How quickly can developed nations transform themselves? How soon will that reinvention also help the billions of other people of this world to achieve a stable, cleaner, healthier, safer life? These are important questions indeed, and they must be answered at the ballot box, in boardrooms, in citizens' assemblies, and in the marketplace.

"Invocation of moral principles is a necessary first step for eliciting virtuous behavior," Jared Diamond writes in *Collapse*, "but that alone is not a sufficient step. Businesses have changed when the public came to expect and require different behavior, to reward businesses for behavior that the public wanted, and to make things difficult for businesses practicing behaviors that the public didn't want. I predict that in the future, just as in the past, changes in public attitudes will be essential for changes in businesses' environmental practices." This is true for governments as well, and for all of their policies. Former U.S. Senator Tim Wirth, leader of the Energy Future Coalition, put it well when he said: "This is not a partisan issue—it is about our future and the kind of world we live in and leave for our children." How we approach global climate change informs our approach to other major world problems— war, poverty, overpopulation, disease, and damage to the diversity of life that supports us all. These are all dangers that threaten humanity, and they create the crisis that E. O. Wilson calls the bottleneck: We are getting into a tight spot, and we must change or we will not get through. "If we get through the bottleneck," he said, "while bringing through as much of the rest of life as we can, for the benefit of all generations, then it will be considered in future centuries, a great accomplishment of this century. . . . I ask, what could be a more noble goal than that?"[87]

EPILOGUE

Mission: Possible

The earth is warming rapidly, and climate change effects are spreading faster than most of us realize. For thousands of years, human civilization has flourished on a hospitable Earth, but that hospitality is waning. Everyone in every nation, and the entire natural world, will be affected.

Fossil fuels, the very power that brought civilization to this point of progress and peril, are responsible for this global change. What comes next could be profoundly negative, rendering hollow the promises of life, human rights, and peace that we hold out to our children and the rest of the world.

We already have the tools to begin changing our energy sources and avoiding waste. Many are now in use; we must do much more. The money, technology, and skills exist in abundance, but we have yet to muster the needed will. Changes must be made at every level of personal life— communal, commercial, national, and international. What's called for is nothing short of a transformation of the world.

Heroes of our new world will emerge based on their contribution to keeping the Earth from getting too hot. They will ignore the usual rules about which nation is "developed" and which is "developing" and will look only at greenhouse emissions. They will realize that the prosperity of rich but heavily polluting nations is no more threatened by change, proportionally, than is that of a country with lower emissions but deep poverty and few financial reserves. The new leaders will know that protecting income, profits, and market share by avoiding action on climate shortsightedly dismisses the possibilities both for disaster and for profitable change. And they will be willing to admit and accept the tremendous difficulty of this transformation.

Let me state the goal clearly: No policy should be promulgated, no program initiated, no alliance sealed, no purchase made, no machine designed or built, no land use permitted, no product introduced,

OPPOSITE Ice cave in a receding glacier in Antarctica. Not long after the photograph was taken, the cave melted away. (JANUARY 2000)

no law passed, no politician elected unless the action is a step forward to reduction and reversal of the effect of greenhouse gases.

Creating energy-conscious urban and industrial infrastructure must begin today, because poor choices will pollute far into the future. No more coal-powered plants until we've tapped every efficiency and renewable energy source in the area. Plants that are built must employ the most effective emissions-reduction technology now available and be compatible with future CO_2 sequestration equipment.

The role for national and international policy is broad. It includes leveling the tax/subsidy playing field for all technologies and promoting those that are most energy efficient; strengthening research and energy reduction programs and expanding them toward renewables; implementing cap-and-trade plans for industrial greenhouse gases; facilitating the large-scale adoption of low-carbon technologies, from lightbulbs to industrial processes; initiating education programs and open communication with scientists; and increasing funds for Earth observation, climate science, and engineering for emissions reductions. Except for the first item, these measures must also be undertaken by corporations and well-endowed charities, without waiting for a change in public policy.

Although the price may appear steep, the damages that human-made climate disruption will otherwise inflict on cities, infrastructure, health, and human life would be far more costly. By contrast, hundreds of millions of dollars may be saved by increasing the efficiency of energy production and use and by reducing waste in the life cycle of goods and buildings. Indeed, this is the place to begin. It is much faster and cheaper to use less energy than to buy it or to reengineer its source.

CREATING A SAFER, CLEANER—AND COOLER—WORLD. This is the direction for change. The world's nations will recognize in global warming a common danger to their people, their cultures, and the Earth that supports them all. The United States, the European Union, China, India, Russia, Japan, Canada, and the rest of the industrialized world must work in concert with the many smaller countries, provinces, states, cities, and corporations already showing the way.

International political and economic relationships will shift to match new alliances and to reflect new regional cooperation. Coalitions of businesses, environmental and social action groups, interfaith groups, international aid groups, local and state governments, foundations, investment groups, and new associations of citizens will organize to address specific climate-related needs that have become obvious all over the world.

With all these positive changes, air and water pollution, smog, smoke, dust, and their detrimental health effects will be greatly reduced. The natural processes and resources that runaway emissions and rising temperatures threaten to disrupt will change less abruptly and will remain healthy and able to support the Earth's population. Many millions of people will enjoy healthier lives.

Is this science fiction? Utopia? *Utopia* means "no place," an imaginary perfection. As James Gustave Speth wrote in *Red Sky at Morning,* "Contrary to the conventional perspective, it is business as usual that is utopian, whereas creating a new consciousness is a pragmatic necessity." The point is, the tools and knowledge we need to bring us a better, cooler, more humane planet are at hand right now. Our heath, security, and well-being depend on our using them.

AFTERWORD

Bill McKibben

In 1989, when I was writing *The End of Nature,* I interviewed an MIT professor named Kerry Emanuel. One of the country's leading hurricane scientists, he'd done some of the early work to suggest that eventually warmer ocean temperatures might increase the intensity of tropical storms. But with the usual reserve of a careful researcher, he wouldn't say more than that.

I thought of that interview this week, when I read Emanuel's latest paper. It showed that hurricanes were already 50 percent more intense than a generation ago—it showed, in short, that we are already seeing effects of climate change far more powerful than even the biggest Cassandras would have predicted just fifteen years ago.

Ditto for the rapid melt of Arctic ice in 2007 and 2008, or the sudden surge of methane pouring out from under the boreal tundra. But we didn't just underestimate the damage. We made another mistake, too: we overestimated the speed with which the world's political systems would react to this greatest of all challenges. Twenty years after the alarms were sounded, we have very little to show for our efforts—only the modest Kyoto treaty, unsigned by the biggest polluter, and due to expire before long with no coherent plan for an extension or replacement. Perhaps this should come as no great surprise; after all, dealing with global warming means weaning the world from fossil fuel, a substance that underwrites its daily operation. Perhaps, as one academic suggested years ago, this is "the public policy problem from hell," one that simply won't be solved in time.

And yet, and yet. There are so many tantalizing signs of the things that could be done.

Some of these are technological. Not super high tech, like the elusive hydrogen car, but off the shelf. Hybrid cars are selling faster than SUVs, and suddenly small houses are chic, McMansions the ugly reminder of a mindless era. Add to these examples lofty insulation, compact fluorescent light bulbs, and so on, and you can go a long way. These are the commodities of the future, and they're being made in the economies that will dominate the future.

More important, the politics of climate change are shifting fast, too. In 2007 a few of us organized two thousand demonstrations around the United States, building support for rapid cuts in emissions—there was a movement waiting to happen, and it shifted the presidential contenders far to the left. Now we're trying to do the same thing globally with 350.org—that being the number, in parts per million CO_2, that the science now tells us is the maximum safe level in the atmosphere.

Reaching that goal will mean building a new world. But it doesn't need to be marked by privation and scarcity. Consider the following statistic, as important as any in the world: Europeans use about half as much energy per capita as Americans. Not because they live in caves; not because they lead deprived lives. But because they lead somewhat more public lives. Their houses are smaller because they share their cities, treating them as extended living rooms. They don't need big cars, or often any car at all, because they've built decent shared systems of transit. And so forth.

One of the key questions—a question that will go a long way toward determining just how hot the planet gets—is whether China and India will take Europe as an example, or look to America. The sooner they start to deflect their trajectories away from ours, the better, because with each passing year the options narrow, and the math gets harder. How many hurricanes will it take, exactly, for the message to start getting through?

ACKNOWLEDGMENTS

This book would not have been possible without help at every point from people who believed in the importance of the issue and in me.

Earth under Fire is an outgrowth of World View of Global Warming, a photo-documentation that I started in March 1999. Two years before that, in Alaska to follow both nature and environmental photo-stories, I had seen 80,000 caribou in the Arctic National Wildlife Refuge and, several days later, the Prudhoe Bay oil fields only 100 miles (160 km) to the west. The incongruity made a powerful impression. Then in May 1998 I read Bill McKibben's *Atlantic Monthly* article about climate change, "A Special Moment in History." In the margin I scrawled "world tour of global warming," not knowing where the idea would lead. It led first to a series of cold calls and emails to scientists and editors. In the years since then (especially in 2000 and 2001, and again in 2004 and 2005, the years when I found the most funding) I have been helped by many hundreds of people around the planet. The following is a recounting of the people and organizations most important in the completion of my project and this book. I am sorry not to be able to acknowledge everyone by name, down to the skillful drivers, staff at field stations, and assistants I have called on to get my work done. To all of them, a humble and heartfelt thank you.

Special thanks go to my son, Cedar Braasch, and his mother, MJ Anderson; my mother, Alice Braasch; and my sister, Peg Strickland, and her husband, Rogers.

Primary funding came from the Wiancko Family Trust through Blue Earth Alliance. Other financial support came from the American Association for the Advancement of Science, with the help of Virginia Stern; the Aveda Environmental Sustainability Program, with the help of Mary Tkach; Tom Campion; Steve Petranek and *Discover* magazine; Jim Motavalli and *E* magazine, Zack Franks; Natural Resources Defense Council; the National Wildlife Federation, with the help of John Nuhn; North American Nature Photography Association; One World Journeys, with the help of Russell Sparkman; and the Packard Foundation/SeaWeb. These funders provided money in the form of grants, assignments, and purchase of prints.

For operational matters, thanks are due to David Kelly, for editorial review; Mary Marlowe, for travel arrangements; Michael McGuire and David McLaughlin, for administration and editing; and Ancil Nance, for website management. In addition I want to thank Elisabeth Weed, my book agent; Phil Borges, Natalie Fobes, Julee Geyer, David Johnson, and Larry Chen of Blue Earth Alliance; the U.N. Rio Conventions staff and calendar designers, particularly Kevin Grose of UNFCCC and James Ramsey of Entico; and the volunteer pilots for SouthWings and Lighthawk, notably Arthur Hussey.

Scientists not only provided information in the field, in interviews, and by phone and email, but many also reviewed sections of the text. Premier acknowledgement goes to those who took the extra time to write essays for this book: Alton Byers, Sylvia Earle, Paul Epstein, Peter Gleick, Thomas Lovejoy, Jonathan Overpeck, Camille Parmesan, and Stephen Schneider; as well as Janica Lane and Cristina Mittermeier, co-authors with Schneider and Earle, respectively. Other scientists offering access, information, and reviews included Jan Aberg, Alcides Ames, Steve Amstrup, Kristina Backstrand, John Bater, Chuck Baxter, Trevor Beebee, Erik Beever, Ed Berg, Andrew Blaustein, Nina Bradley, Raymond Bradley, Kent Bransford, Dan Breneman, Julie Brigham-Grette, Jerry Brown, Terry Callaghan, Geoff Carroll, Scott Chambers, Terry Chapin, M. H. Khan Chowdhury and other members of NOAMI, John Christy, Luke Copland, Humphrey Crick, Patrick Crill, Lisa Crozier, Bill Curry, Ruth Curry, Andrew Derocher, Debra DeRosier, George Divoky, Eugene Domack, Terry Done, Robert Dunbar, Dan Fagre, Andrew Fountain, Bill Fraser, Hector Galbraith, John Gammon, Shari Fox Gearheard, Mickey Glantz, Thomas Goreau, Andrew Gosler, Michael Gottfried, Nicholas Graham, Brad Griffith, Hermann Gucinski, Roel Hammerschlag, Lara Hansen, John Harte, Stephen Hawkins, Kevin Helmle, Kenneth Hewitt, Jos Hill, Ove Hoegh-Guldberg, Marcel Holyoak, Malcolm Hughes, David Inouye, Torre Jorgenson, Glenn Patrick Juday, Jeff Kargel, Thomas Karl, Georg Kaser, Eric Kasischke, Boone Kauffman, Lloyd Keigwin, Joanie Kleypas, Bill Krabill, Jack Kruse, Daniel Lashof, Richard LaVal, Paul LeFebvre, Corinne LeQuéré, Amy Leventer, Amory Lovins, Brian Luckman, Jane Lubchenco, Michael MacCracken, Michael Mann, Adam Markham, Paul Marshall, Karen Masters, Mark F. Meier, Annette Menzel, Jeff Miller, Steve Montgomery, Bob Moseley, Ellen Mosley-Thompson, Nalini Nadkarni, Ron Neilson, Elliott Norse, Walt Oechel, Tom Osterkamp, Steve Palumbi, Jonathan Patz, Harald Pauli, Mark Peck, Mauri Pelto, Bill Peterson, Bruce Peterson, David Pimentel, Stuart Pimm, John Porter, Alan Pounds, Jeff Price, Kenneth Raffa, Stefan Ramstorf, Peter Raven, Anthony Richardson, Eric Rignot, Vladimir Romanovsky, Terry Root, Cynthia Rosenzweig, Raphael Sagarin, Scott Schliebe, Gavin Schmidt, Mark Serreze, Gus Shaver, Justine Shaw, Glenn Sheehan, Stan Shetler, Christopher Shuman, Michael Singer, Heinz Slupetzky, Anthony Socci, Robert Socolow, Alan Southward, Tim Sparks, Konrad Steffen, Nathan Stephenson, Paul Strode, Matthew Sturm, William Sydeman, Chris Thomas, Bob Thomas, Lonnie Thompson, John Toole, Ramzi Touchan, Kevin Trenberth, Craig Tweedie, Johan Varekamp, Carden Wallace, Harold Wanless, Spencer Weart, Gunter Weller, Rick Wessels, Ken Whitten, Clive Wilkinson, Steve Williams, George Woodwell, and five anonymous reviewers. I also thank the scientists of NASA and NOAA; Nikita Lopoukhine, IUCN Commission on Protected Areas, for his comments; and Richard Leakey for the opportunity to attend the Life Matters World Environment Forum.

For additional advice, support, and information I wish to acknowledge Peter Arnold, Nancy Baron, Niki Barrie, Spencer Beebe, Tom Bender, Nancy and Dennis Biasi, Connie Bransilver, Colin Brown, Jessica Brown, Wil Burns, Bobbi Baker Burrows, Woodfin Camp, Lynne Cherry, Josh Cherwin, Aimee Christensen, Jonathan Clough, Ana Unruh Cohen, Ed Cookman, Rosamund Kidman Cox, Wade Davis, Chuck Dayton, Patty Debenham, Jan DeBlieu, Ann Dilworth, Mark Drew, Mark Edwards, Jerry Ellis, Josef Essl, Seth Fields, Matthew Follette, David Friend, Kathleen Frith, Frances Gatz, Ross Gelbspan, Larry Gibson, Nathan Glasgow, Patty Glick, Eban Goodstein, David Gregg, Anne Guilfoyle, Wolf Guindon, Joel Halioua, Patrick Hamilton, Archie Harders, David Hawkins, Denis Hayes, Judith Helfland, Tim Hermach, Tim Herzog, Martin Hiller, Kim Hubbard, Jon Isham, Jane Kinne, Kelly Knee, Jay Kravitz, Kalee Kreider, Tom LaBerge, Bart Lewis, Li Moxuan, Stephanie Long, Barry Lopez, Elizabeth Losey, David Lyman, Mark Lynas, Ken Margolis, Elizabeth May, Fred Mays, Patrick Mazza, Bill McKibben, Scott McKiernan, Cristina Mittermeier, Jennifer Morgan, Gordon Murdock, Elliott Negin, Robert O'Connor, Lance Olson, Peter Otis, Lello Piazza, Joanna Priestley, Frances Raskin, Doug Ray, Richard Register, Ellen and Bob Reynolds, Dan Ritzman, Patricio Robles Gil, Hans Rosling, John Ryan, Melissa Ryan, David Sandalow, Cristina Scalet, Claudine Schneider, John Schoen, Kassie Siegel, James Gustave Speth, John Stanton, Miriam Stein, Vivian Stockman, Mike Tidwell, Sandy Tolan, Doug and Kristine Tompkins, John Topping, Frank Tursi, Stefan Wagner, Sheila Watt-Cloutier, Dennis Wiancko, Mary Wildfire, Deborah Williams, Ike Williams, Joshua Wolfe, and the many other friends, fellow photographers, and helpful strangers who offered support. I thank the organizations that have invited me to speak at meetings and seminars. This book enjoys the support of the International League of Conservation Photographers.

Beyond those recognized above, I would like to acknowledge the following colleagues and friends who so generously aided my work by providing invaluable contacts and helping with logistics. In Antarctica, the staff of the National Science Foundation's Office of Polar Programs, especially Scott Borg, Faye Korsmo, Polly Penhale, and Peter West; the crews of the *R/V Palmer* and *Gould* and personnel of Palmer Station; and Ken Doggett, Steve Dunbar, Jim LoScalzo, Ron Naveen, and Charles Petit. In the Arctic, the personnel of the National Science Foundation's Toolik Lake Station; the staff of the Abisko Scientific Research Station; Keith Koehler and the crew of the P-3 aircraft from NASA Goddard Space Flight Center at Wallops Island; Jim Campbell and the men of the Marsh Fork; the leaders and elders of the villages of Pangnirtung, Nunavut, and Shishmaref, Alaska; Patricia Anderson, Graham Ashford, Walt Audi, Patricia Cochran, Jim Helmericks, John Houston, Jeanne Mike, Debbie and Dennis Miller, Pam Miller, Bruce Molnia, Eva Sowdluapik, Ken Tape, Sylvia Ward, Tony Weyiouanna Sr., and Carla Willetto. In England, Nigel Hepper, Michele Taylor, Chris Thain, David Walker, and Ian Woiwod. In Peru, Edwin Bernbaum, Nilda Callañaupa, Bryan Mark, Celso Jaimes Quispe, Jorge Recharte, Ankur Tohan, and Miriam Torres. In Pacific locations, the government and ministerial staff of Tuvalu; Great Barrier Reef Marine Protected Area staff; Ant Backer, Kilateli Epu, Koin Etuati, Kim Friedman, Gregor Hodgson, Julie Jones, Chalapan Kaluwin, Paani Laupepa, Alice Leney, Francois Martel, Laurence McCook, John Novis, Siuila Toloa, Gareth Walton, and James Williams. In China, Chen Qi, Barbara Finamore, Li Moxuan, Liu Zhi, Qian JingJing, Ru Jiang, Wang Yongchen, and Zhao Ang. In Bangladesh, Enan Bhai, Enam Ul Haque, Anwar Islam, Mamunul Khan, M. A. Mohit and his family on Bhola Island, and Sally Walker. In Kenya, Candace and Bob Buzzard, Steve Jackson, Nick Nuttall, Peter Smerdon, and staff of UNFCCC and UNEP.

NGOs and organizations that were especially helpful include the Alaska Coalition, the Alaska Wilderness League, the Aldo Leopold Foundation, Catholic Earthcare Australia, the Center for Health and the Global Environment, Clean Air–Cool Planet, the Climate Institute, Climate Solutions, Conservation International, Earthjustice, Ecotrust, E & E Publishing, Environmental Defense, GreenCities, the Greenhouse Network, Greenpeace, *Grist* magazine, the Heinz Center for Science, Economics and the Environment, the Institute for Energy and Environmental Research, the International Institute for Sustainable Development, Lighthawk, the Mountain Institute, the National Audubon Society, the National Environmental Trust, the Native Forest Council, the Natural Resources Defense Council, the Nature Conservancy, NatureServe, the Nieman Foundation, the Northern Alaska Environmental Center, Physicians for Social Responsibility, SEAWeb, the Sierra Club, the Society of Environmental Journalists, SouthWings, the World Resources Institute, the World Wildlife Fund, UNEP/GRID, the Union of Concerned Scientists, the Western Atlantic Shorebird Association, and the many other organizations whose websites have supplied valuable information and that have displayed my photographs.

At University of California Press, thanks go to Blake Edgar, my patient and insightful editor; editorial and production staff of Rose Vekony, Dore Brown, Chalon Emmons, and Matthew Winfield; designer Nicole Hayward; and freelance copy editors Anne Canright and Sheila Berg. I thank also the scientists and agency employees who provided photographs, and who are credited in the captions.

I deeply apologize if I have neglected to mention anyone.

Any errors in fact or omission in this book are mine alone. I planned, photographed, researched, wrote, and raised money for this project personally. My regular photography income also contributed, and I thank the many magazines, book publishers, and exhibit spaces that have paid to use my work.

My project, World View of Global Warming, continues with more documentary photography, educational publications, and presentations. Donations are accepted through Blue Earth Alliance, PO Box 94388, Seattle WA 98124-6688, a 501(c)3 organization. Email info@blueearth.org for information.

The goal of this work has been to witness and provide rare documentation of climate changes in the world, yet traveling for it has created a lot of greenhouse gases; there is no way around that. I have taken the step of mitigating my air travel and energy use through the Carbon Fund and Native Energy. I own a house of less than 1,200 square feet, have one child (now at college), use PG&E's Healthy Habitat source of electricity, and ride bicycles, not power-toys. Still, being an American means I use a lot of energy to support myself and my community. However, I learned a lot about low-impact energy during my travels and trust that it will be the future of my family and all who read this book.

CONTRIBUTORS

ALTON C. BYERS is director of research and education programs for the Mountain Institute in Washington DC. He is coeditor of the forthcoming book *Mountains and People*.

SYLVIA A. EARLE, former chief scientist for the National Oceanic and Atmospheric Administration, is an oceanographer and an explorer in residence at the National Geographic Society. She is the author of *Sea Change: A Message of the Oceans.*

PAUL R. EPSTEIN, MD, MPH, is associate director of the Center for Health and the Global Environment at Harvard Medical School. He is coeditor of *Ecosystem Health: Principles and Practice.*

PETER H. GLEICK is cofounder and president of the Pacific Institute for Studies in Development, Environment, and Security in Oakland, California. He is the author of *The World's Water 2006–2007* and the editor of *Water in Crisis.*

JANICA LANE is a student in the Graduate School of Business and a former research fellow of the Freeman Spogli Institute for International Studies at Stanford University.

THOMAS E. LOVEJOY, a conservation biologist, is president of the H. John Heinz III Center for Science, Economics and the Environment in Washington DC. He is coeditor of *Climate Change and Biodiversity.*

BILL McKIBBEN is a scholar in residence at Middlebury College and the author of nine books, including *The End of Nature* and *Deep Economy.*

CRISTINA G. MITTERMEIER is a marine biologist and executive director of the International League of Conservation Photographers.

JONATHAN OVERPECK is director of the Institute for the Study of Planet Earth and professor of geosciences at the University of Arizona. He is the recipient of the U.S. Department of Commerce Gold Medal and the American Meteorological Society's Walter Orr Roberts Award.

CAMILLE PARMESAN is associate professor in the Section of Integrative Biology at the University of Texas, Austin. She was a lead author of the Third Assessment Report of the Intergovernmental Panel on Climate Change.

STEPHEN H. SCHNEIDER, a climatologist, is a senior fellow at the Stanford Institute for International Studies and a professor in the Department of Biological Sciences at Stanford University. He is the author of *Laboratory Earth* and coeditor of *Wildlife Responses to Climate Change* and *Climate Change Policy.*

NOTES

In the spirit of the scientific work on which this book is based, I have chosen to cite all of my major sources, and to be clear when general information is taken from public media sources. The most recent reference for this book was accessed on 2 October 2008. Readers are urged to search out the most recent developments. Of particular interest will be the 2007 Intergovernmental Panel on Climate Change assessments, both the summaries for policymakers and the more detailed technical summaries, forthcoming throughout 2007. I have included, to the extent possible, the underlying research on which these assessments are based.

CHAPTER 1

1 For Antarctic changes, see David C. Vaughan et al., "Recent Rapid Regional Climate Warming on the Antarctic Peninsula," *Climatic Change* 60 (Oct. 2003); also see J. C. King, "Antarctic Peninsula Climate Variability and Its Causes as Revealed by Instrumental Records," in *Antarctic Peninsula Climate Variability: A Historical and Paleoenvironmental Perspective,* Antarctic Research Series, vol. 79, ed. Eugene W. Domack et al. (Washington, DC: American Geophysical Union, 2003). Summaries of the current knowledge of climate change, compiled and written by the Intergovernmental Panel on Climate Change, may be obtained at www.ipcc.ch. See also Naomi Oreskes, "The Scientific Consensus on Climate Change," *Science* 306 (3 Dec. 2004).

2 Curt Davis et al., "Snowfall-Driven Growth in East Antarctic Ice Sheet Mitigates Recent Sea-level Rise," *Science* 308 (24 June 2005); Isavella Velicogna and John Wahr, "Measure-

ments of Time-Variable Gravity Show Mass Loss in Antarctica," *Science* 311 (24 Mar. 2006); "NASA Mission Detects Significant Antarctic Ice Mass Loss," 2 Mar. 2006, www.jpl.nasa.gov/news/news.cfm?release=2006–028; Andrew J. Monaghan et al., "Insignificant Change in Antarctic Snowfall Since the International Geophysical Year," *Science* 313 (11 Aug. 2006); David P. Schneider et al., "Antarctic Temperatures over the Past Two Centuries from Ice Cores," *Geophysical Research Letters* 33 (30 Aug. 2006); Eric Steig, "Is Antarctic Climate Changing?," 25 Aug. 2006, www.realclimate .org/index.php/archives/2006/08/antarctica-snowfall; Cecelia Bitz, "Polar Amplification," 2 Jan. 2006, www.realclimate .org/index.php/archives/2006/01/polar-amplification; Robert Bindschadler, NASA Goddard Space Flight Center, Greenbelt, MD, speaking at Washington Summit on Climate Stabilization, 19 Sept. 2006.

3 Eric Rignot, email, 9 Jan. 2006; Vaughn et al., "Recent Rapid Regional Climate Warming"; King, "Antarctic Peninsula Climate Variability"; Raymond C. Smith et al., "Marine Ecosystem Sensitivity to Climate Change," *Bioscience* 49, no. 5 (1999); National Snow and Ice Data Center (NSIDC), "State of the Cryosphere: Ice Shelves," http://nsidc.org/iceshelves.

4 Data on Arctic temperatures is from F. S. Chapin et al., "Role of Land-Surface Changes in Arctic Summer Warming," *Science* 310 (28 Oct. 2005), and Glenn Patrick Juday, Professor of Forest Ecology, University of Alaska, email, 15 Oct. 2005. Other information throughout this chapter is from Larry D. Hinzman et al., "Evidence and Implications of Recent Cli-

mate Change in Northern Alaska and Other Arctic Regions," *Climatic Change* 72 (Nov. 2005); Arctic Climate Impact Assessment (ACIA), *Impacts of a Warming Arctic* (Cambridge: Cambridge University Press, 2004); J. Richter-Menge et al., "State of the Arctic Report," NOAA OAR Special Report, NOAA/OAR/PMEL (October 2006). Also see Matthew Sturm, Donald Perovich, and Mark Serreze, "Meltdown in the North," *Scientific American,* Oct. 2003.

5 Christopher A. Shuman, NASA, email, 17 Jan. 2007. For information on the 2002 Larsen Ice Shelf breakup, see http://nsidc.org/iceshelves/larsenb2002.

6 Eugene Domack et al., "Stability of the Larsen B Ice Shelf on the Antarctic Peninsula during the Holocene Epoch," *Nature* 436 (4 Aug. 2005); and Eugene Domack et al., "Marine Sedimentary Record of Natural Environmental Variability and Recent Warming in the Antarctic Peninsula," in Domack et al., eds., *Antarctic Peninsula Climate Variability.*

7 Interview with Eugene Domack, Apr. 1999; Domack et al., eds., *Antarctic Peninsula Climate Variability;* Ellen Mosley-Thompson in L. G. Thompson et al., "Climate since A.D. 1510 on Dyer Plateau, Antarctic Peninsula: Evidence for Recent Climate Change," *Annals of Glaciology* 20 (1994).

8 Scambos is quoted in American Geophysical Union (AGU) press release no. 04–33, 21 Sept. 2004. Two other studies— Eric Rignot et al., "Accelerated Ice Discharge from the Antarctic Peninsula Following the Collapse of the Larsen B Ice Shelf," and Ted Scambos et al., "Glacier Acceleration and Thinning after Ice Shelf Collapse in the Larsen B Embayment, Antarctica," both in *Geophysical Research Letters* 31 (22 Sept. 2004)—used airborne GPS (global positioning system) mapping, satellite radar, and remote sensing data to track the same five glaciers' surges toward the sea; both noted dramatic thinning of the ice, another indication of fast movement. See also Hernan de Angelis and Pedro Skvarca, "Glacier Surge after Ice Shelf Collapse," *Science* 299 (7 Mar. 2003).

9 Eric Rignot et al., "Recent Ice Loss from the Fleming and Other Glaciers, Wordie Bay, West Antarctic Peninsula," *Geophysical Research Letters* 32 (14 Apr. 2005); A. J. Cook et al., "Retreating Glacier Fronts on the Antarctic Peninsula over the Past Half-Century," *Science* 308 (22 Apr. 2005); Rignot, email, 9 Jan. 2006; Robert Thomas, email, 16 Nov. 2005.

10 R. Thomas et al., "Accelerated Sea-Level Rise from West Antarctica," *Science* 306 (8 Oct. 2004); Robert Bindschadler, "Hitting the Ice Sheets Where It Hurts," *Science* 311 (25 Mar. 2006); Andrew Shepherd et al., "Warm Ocean Is Eroding West Antarctic Ice Sheet," *Geophysical Research Letters* 31 (9 Dec. 2004); Helen Amanda Fricker et al., "An Active Subglacial Water System in West Antarctica Mapped from Space," *Science* 315 (16 Mar. 2007); Eric Rignot and Robert

Bindschadler, Washington Summit on Climate Stabilization, Sept. 18–21, 2006. Also see Michael Oppenheimer, "Ice Sheets and Sea Level Rise: Model Failure Is the Key Issue," 26 June 2006, www.realclimate.org/index.php/archives/2006/06/ice-sheets-and-sea-level-rise-model-failure-is-the-key-issue.

11 Smith et al., "Marine Ecosystem Sensitivity to Climate Change"; interviews with Bill Fraser and other scientists at Palmer Station, 1999 and 2000.

12 Jared Diamond, *Collapse: How Societies Choose to Fail or Succeed* (New York: Viking-Penguin, 2005), 212.

13 Jonathan Overpeck et al., "Arctic Environmental Change of the Last Four Centuries," *Science* 278 (14 Nov. 1997); M. C. Serreze et al., "Observational Evidence of Recent Change in the Northern High-Latitude Environment," *Climatic Change* 4 (July 2000); Serreze et al., "A Record Minimum Arctic Sea Ice Extent and Area in 2002," *Geophysical Research Letters* 30 (5 Feb. 2003); and Konrad Steffen and Russell Huff, "Greenland Melt Extent, 2005," 28 Sept. 2005, http://cires.colorado.edu/science/groups/steffen/greenland/melt2005. For glacier measurements, see next note.

14 Information on the NASA Greenland mapping project can be found at www.nasa.gov/centers/goddard/earthandsun/thinningice.html. See also W. Krabill et al., "Greenland Ice Sheet: High-Elevation Balance and Peripheral Thinning," and R. Thomas et al., "Mass Balance of the Greenland Ice Sheet at High Elevations," both in *Science* 289 (21 July 2000).

15 Robert Thomas, email, Nov. 2005, based on papers in press. Also see Eric Rignot and Pannir Kanagaratnam, "Changes in the Velocity Structure of the Greenland Ice Sheet," *Science* 311 (17 Feb. 2006), which discusses another east Greenland glacier, the Helheim, that also appears to be rapidly flowing and thinning; William Krabill et al., "Greenland Ice Sheet: Increased Coastal Thinning," *Geophysical Research Letters* 31 (28 Dec. 2004); Ian Joughin et al., "Large Fluctuation in Speed on Greenland's Jakobshavn Isbrae Glacier," *Nature* 432 (2 Dec. 2004); Robert Thomas et al., "Investigation of Surface Melting and Dynamic Thinning on Jakobshavn Isbrae, Greenland," *Journal of Glaciology* 49 (2003).

16 Information on ice coring is from www.ncdc.noaa.gov/paleo/icecore/greenland/summit/document/gripinfo.htm and www.agu.org/revgeophys/mayews01/node2.html; Younger Dryas information from Richard Alley, *The Two-Mile Time Machine: Ice Cores, Abrupt Climate Change, and Our Future* (Princeton: Princeton University Press, 2000), 4, 110–22. Longer cycles that appear to cause ice ages to come and go can be seen farther back in time beyond the ice core record, in sediment cores from the Atlantic Ocean. There are many cycles, and abrupt changes seem to be the norm. Alley says that the current 11,000-year period of comparatively stable climate is one of the longest on record.

17 Andrew Shepherd and Duncan Wingham, "Recent Sea Level Contributions of the Antarctic and Greenland Ice Sheets," and Martin Truffer and Mark Fahnestock, "Rethinking Ice Sheet Time Scales," both in *Science* 315 (16 Mar. 2007); Ian Howat et al., "Rapid Changes in Ice Discharge from Greenland Outlet Glaciers," *Science Express,* 8 Feb. 2007; Howat, email, 14 Feb. 2007; H. Jay Zwally et al., "Surface Melt–Induced Acceleration of Greenland Ice-Sheet Flow," *Science* 297 (12 July 2002); Mark B. Dyurgerov and Jason E. Box, "Arctic Glaciers' Dynamic Characteristics and Implicit Sensitivity: Global Sea Level Implications," paper presented at the 36th Annual Arctic Workshop, Institute of Arctic and Alpine Research, 16 Mar. 2006; email exchange with Thomas (previously cited).

18 Email exchanges with Thomas and Rignot (previously cited); and Thomas, email, 3 Oct. 2006; A. F. Glazovskiy and M. B. Dyurgerov, "Glacier Contribution to the Arctic Ocean," *Eos* Supplement 87 (11 Dec 2006).

19 T. Toniazzo et al., "Climatic Impact of a Greenland Deglaciation and Its Possible Irreversibility," *Journal of Climate* 17 (Jan. 2004); Richard Alley et al., "Ice-Sheet and Sea-Level Changes," *Science* 310 (21 Oct. 2005); Julian Dowdesnell, "The Greenland Ice Sheet and Global Sea Level Rise," *Science* 311 (17 Feb. 2006); J. T. Overpeck et al., "Paleoclimatic Evidence for Future Ice Sheet Instability and Rapid Sea Level Rise," *Science* 311 (24 Mar. 2006). An overview of Greenland ice science may be found in Quirin Schiermeier, "A Rising Tide," *Nature* 428, 11 Mar. 2004. See also Donald Kennedy and Brooks Hanson, "Ice and History," *Science* 311 (24 Mar. 2006).

20 D. A. Rothrock et al., "Thinning of the Arctic Sea Ice Cover," *Geophysical Research Letters* 26 (1 Dec. 1999). For figures and analysis for 2005, see http://nsidc.org/news/press/20050928_trendscontinue.html; for 2007, http://nsidc.org/news/press/2007_seaiceminimum/20071001_pressrelease.html.

21 Serreze quotes in this discussion are from an interview in Salt Lake City, 3 Mar. 2006; emails, 30 Jan. 2006; and telephone, 2 Oct. 2006. Josefio Comiso, "Abrupt Decline in the Arctic Winter Sea Ice Cover," *Geophysical Research Letters* 33 (30 Sept. 2006); idem, "A Rapidly Declining Perennial Sea Ice Cover in the Arctic," *Geophysical Research Letters* 29 (18 Oct. 2002); Mark C. Serreze et al., "Perspectives on the Arctic's Shrinking Sea-Ice Cover," *Science* 315 (16 Mar. 2007); http://nsidc.org/arcticseaicenews/2008/040708.html.

22 Mark Serreze and Jennifer Francis, "The Arctic on the Fast Track of Change," *Weather* 61 (Mar. 2006); Serreze interview, 3 Mar. 2006.

23 "Giant Arctic Ice Shelf Cracks Up," *Geophysical Institute Quarterly* (University of Alaska, Fairbanks) 18, no. 4 (2004); Nathan Vander Klippe, "Warming Threatens Nunavut Ice Shelf," CanWest News Service, available at www.sciencedaily.com/releases/2003/09/030923065136.htm; Canadian Ice Service, "The Calving of the Ayles Ice Shelf," 27 Jan. 2007, at http://iceglaces.ec.gc.ca; Bruce Peterson et al., "Increasing River Discharge to the Arctic Ocean," *Science* 298 (13 Dec. 2002).

24 T. E. Osterkamp, "The Recent Warming of Permafrost in Alaska," *Global and Planetary Change* 49, nos. 3–4 (Dec. 2005); V. E. Romanovsky, "How Rapidly Is Permafrost Changing and What Are the Impacts of These Changes?" Arctic Theme Page, National Atmospheric and Oceanic Administration (NOAA), www.arctic.noaa.gov/essay_romanovsky.html. Generalization about world permafrost is from AGU meeting reports, Dec. 2004.

25 Osterkamp, email, Jan. 2006; Torre Jorgenson et al., "Abrupt Increase in Permafrost Degradation in Arctic Alaska," *Geophysical Research Letters* 33 (24 Jan. 2006); L. C. Smith et al., "Disappearing Arctic Lakes," *Science* 308 (3 June 2005); Sergey A. Zimov et al., "Permafrost and the Global Carbon Budget," *Science* 312 (16 June 2006); Bob Henson, "The Big Thaw: New CCSM Runs Predict More Trouble for Arctic Sea Ice, Permafrost," www.ucar.edu/communications/quarterly/winter0506/permafrost.jsp; Hinzman et al., "Evidence and Implications of Recent Climate Change."

26 Jonathan Overpeck et al., "Arctic System on Trajectory to New, Seasonally Ice-Free State," *Transactions of the American Geophysical Union,* 23 Aug. 2005; also see "Trouble in Polar Paradise," section of "News and Reviews," *Science* 297 (30 Aug. 2002); and the Arctic Council's *Impacts of a Warming Arctic,* available at www.acia.uaf.edu or in book form from Cambridge University Press (Feb. 2005).

27 Wallace Broecker, "The Biggest Chill," *Natural History* 96 (Oct. 1987); and idem, "What If the Conveyor Were to Shut Down? Reflections on a Possible Outcome of the Great Global Experiment," *GSA Today* 9 (Jan. 1999). Scientists often refer to the "conveyor" currents as the Meridional Overturning Circulation (MOC) and the Thermohaline Circulation; Alley, *Two-Mile Time Machine,* 112ff. A detailed history of this discovery of rapid climate change can be found in Spencer Weart, *The Discovery of Global Warming* (Cambridge, MA: Harvard University Press, 2003); see also his website at www.aip.org/history/climate/rapid.htm.

28 Jerry McManus et al., "Collapse and Rapid Resumption of Atlantic Meridional Circulation Linked to Deglacial Climate Changes," *Nature* 428 (22 Apr. 2004).

29 Ruth Curry, interview aboard the *Oceanus,* 20 Oct. 2002. Also see Robert B. Gagosian, "Triggering Abrupt Climate Change," www.whoi.edu/home/about/whatsnew/

abruptclimate.html; and Peter Schwartz and Doug Randall, "An Abrupt Climate Change Scenario and Its Implications for United States National Security," October 2003, www.environmentaldefense.org/documents/3566_AbruptClimateChange.pdf.

30 See R. Curry et al., "A Change in the Freshwater Balance of the Atlantic Ocean over the Past Four Decades," *Nature* 426 (18/25 Dec. 2003); Bob Dickson et al., "Rapid Freshening of the Deep North Atlantic Ocean over the Past Four Decades," *Nature* 416 (25 Apr. 2002); Ruth Curry and Cecilie Mauritzen, "Dilution of the Northern Atlantic Ocean in Recent Decades," *Science* 308 (17 June 2005).

31 Harry L. Bryden, Hannah R. Longworth, and Stuart A. Cunningham, "Slowing of the Atlantic Meridional Overturning Circulation at 25°N," *Nature* 438 (1 Dec. 2005); Gavin Schmidt, "Ocean Circulation: New Evidence (Yes), Slowdown (No)," 31 Oct. 2006, at Real Climate, www.realclimate.org/index.php/archives/2006/10/ocean-circulation-new-evidence-yes-slowdown-no; Bill Curry, Woods Hole Oceanographic Institution, email 24 Jan. 2007; Ruth Curry, personal communication, 7 Jan. 2006; Bruce J. Peterson et al., "Trajectory Shifts in the Arctic and Subarctic Freshwater Cycle," *Science* 313 (25 Aug. 2006). Odds were first presented by Michael Schlesinger at a conference on climate change held in Exeter, England, 1–3 Feb. 2005; see IPCC, "Climate Change 2007: The Physical Science Basis. Summary for Policymakers," 2 Feb. 2007, available at www.ipcc.ch.

32 EPICA Community Members, "One-to-One Coupling of Glacial Climate Variability in Greenland and Antarctica," *Nature* 444 (9 Nov. 2006); Stefan Rahmstorf, "Revealed: Secrets of Abrupt Climate Shifts," at www.realclimate.org/index.php/archives/2006/11/revealed-secrets-of-abrupt-climate-shifts.

33 World Glacier Inventory, at http://nsidc.org/data/g01130.html; Global Land Ice Measurements from Space (GLIMS), www.glims.org; Wilfried Haeberli and Martin Hoelzle, "Alpine Glaciers as a Climate Proxy and as a Prominent Climate Impact," available at www.zamg.ac.at/ALP-IMP/downloads/session_haeberli.pdf.

34 Mark Dyurgerov and Mark Meier, "Glaciers and the Changing Earth System: A 2004 Snapshot," Occasional Paper No. 58, Institute of Arctic and Alpine Research (INSTAAR), Boulder, CO, 2005; Mark Dyurgerov, "Mass Balance of Mountain and Sub-Polar Glaciers Outside the Greenland and Antarctic Ice Sheets," Supplement to Occasional Paper No. 55, INSTAAR, Boulder, CO, 2005; J. Oerlemans, "Extracting a Climate Signal from 169 Glacier Records," *Science* 308 (29 Apr. 2005); World Glacier Monitoring Service, www.wgms.ch; the "twice as fast" estimate is from Anthony A. Arendt et al., "Rapid Wastage of Alaska Glaciers and Their Contribu-

tion to Rising Sea Level," *Science* 297 (19 July 2002). Individual glaciers show variability from decade to decade due to volcanic eruption cooling effects, short-term weather cycles, and changes in moisture sources. There are only a few known exceptions to the trend of glacier mass wastage: a few small glaciers in Norway and New Zealand and some on Mt. Shasta in California, due to local precipitation conditions; a glacier growing in the crater of Mt. St. Helens in Washington, because the location is new since the 1980 eruption; a few glaciers in Alaska such as the Hubbard, because of idiosyncrasies in their basins; and some glaciers in the Himalayas, reflecting that region's wide topographical and weather variety. Although a few glaciers appear to be growing longer, they are also rapidly thinning and losing mass; that is, they have less ice, and mass is what counts in glacier measurements. See http://nsidc.org/sotc/glacier_balance.html.

35 Roger G. Barry, "Assessing Global Glacier Recession: Results of the Workshop," in *Papers and Recommendations: Workshop on Assessing Global Glacier Recession,* Glaciological Data Report GD32, National Snow and Ice Data Center/World Data Center for Glaciology, Boulder, CO, Dec. 2003. Glacier lake outbursts (and glacier shrinkage) are discussed in "Asia's Water Security under Threat," www.unep.org/Documents.Multilingual/Default.asp?DocumentID=452&ArticleID=4916&l=en; Dyurgerov and Meier, "Glaciers and the Changing Earth System." See also Megan Sever, "Melting Glaciers Reveal Ancient Bodies," *GEOTimes,* Apr. 2005; and Johan Reinhard in various *National Geographic* issues. Some fieldwork indicates that heightened volcanic activity occurred during past periods when glaciers retreated; Sharon Begley, "How Melting Glaciers Alter Earth's Surface, Spur Quakes, Volcanoes," *Wall Street Journal,* 9 June 2006.

36 U.N. Food and Agriculture Organization (FAO), "Besieged Mountain Ecosystems Start to Turn Off the Tap," www.fao.org/english/newsroom/news/2002/9881-en.html; T. P. Barnett, J. C. Adam, and D. P. Lettenmaier, "Potential Impacts of a Warming Climate on Water Availability in Snow-Dominated Regions," *Nature* 438 (17 Nov. 2005); Anne Coudrain et al., "Glacier Shrinkage in the Andes and Consequences for Water Resources," *Hydrological Cycles Journal* 50 (Dec. 2005).

37 Dorothy Hall et al., "Changes in the Pasterze Glacier as Measured from the Ground and Space," 58th Eastern Snow Conference, Ottawa, 2001, at www.easternsnow.org/proceedings/2001/Hall1.pdf.

38 Frank Paul et al., "Rapid Disintegration of Alpine Glaciers Observed with Satellite Data," *Geophysical Research Letters* 31 (12 Nov. 2004); "Melting Swiss Glaciers Threaten Alps," *World Environment News,* 16 Nov. 2004, at www.planetark.com/

avantogo/dailynewsstory.cfm/newsid/28158/story.htm; U.N. Environment Programme (UNEP), "Impacts of Summer 2003 Heat Wave in Europe," Mar. 2004, available at www.grid.unep.ch/product/publication/download/ew_heat_wave.en.pdf; Quirin Schiermeier, "Alpine Thaw Breaks Ice over Permafrost's Role," *Nature* 424 (14 Aug. 2003); "Winter Tourism on Thin Ice," *Swiss News,* Aug. 2006; Michael Zemp et al., "Alpine Glaciers to Disappear within Decades?" *Geophysical Researcthrswifh Letters* 33 (15 July 2006).

39 Johannes Koch, John Clague, and Gerald Osborn, "Environmental Change in Garibaldi Provincial Park, Southern Coast Mountains, British Columbia," *Geoscience Canada,* 9 Jan. 2004; Dan Fagre, interview, Sept. 2000; Hassan Basagic and Andrew Fountain, "Glaciers and Glacier Change of the Sierra Nevada, California, USA," poster presented at MtnClim 2006 research conference, Mt. Hood, OR, 19–22 Sept. 2006, abstracts available at www.fe.fed.us/psw/mtnclim. Also see Andrew Fountain et al., "Glacier Response in the American West to Climate Change During the Past Century," poster at MtnClim 2006; Brian Luckman and Trudy Kavanagh, "Recent Environmental Changes in the Canadian Rockies," *Mountain Science Highlights,* no. 18, at www.forestry.ubc.ca/alpine/highlights; and Fagre's website, with links, at www.nrmsc.usgs.gov/staff/fagre.html. I also received information about Cascade Range glacier recession from Robert M. Krimmel, USGS-ICP, Tacoma, WA; Andrew Fountain, Portland State University, Portland, OR; and Mauri Pelto, Nichols College, Dudley, MA. Pelto's long-term studies of Cascade Range ice may be viewed at www.nichols.edu/departments/Glacier/nccent.htm. Another view of glacier oscillation is Don Easterbrook, "Causes and Effects of Late Pleistocene, Abrupt, Global, Climate Changes and Global Warming," Paper 15–13, Geological Society of America Abstracts 37, no. 7 (2005), 41.

40 Philip Mote et al., "Declining Mountain Snowpack in Western North America," *Bulletin of the American Meteorological Society,* Jan. 2005; Robert F. Service, "As the West Goes Dry," *Science* 303 (20 Feb. 2004); Gregory Pederson et al., "Decadal-Scale Climate Drivers for Glacial Dynamics in Glacier National Park," *Geophysical Research Letters* 31 (17 June 2004); Alan F. Hamlet et al., "An Overview of 20th-Century Warming and Climate Variability in the Western U.S.," presentation at MtcClim 2006.

41 Interview with Alcides Ames, Huaraz, Peru, July 1999. Also see "Small Glaciers of the Andes May Vanish in 10–15 Years," 17 Jan. 2001, http://unisci.com/stories/20011/0117013.htm, about studies conducted on the Chacaltaya Glacier in Bolivia and the Antizana Glacier in Ecuador; and B. Francou et al., "Glacier Evolution in the Tropical Andes during the Last Decades of the 20th Century: Chacaltaya, Bolivia, and Antizana, Ecuador," *Ambio* 29, no. 7 (Nov. 2000). Raymond Bradley et al., "Threats to Water Supplies in the Tropical Andes," *Science* 312 (23 June 2006); Coudrain, "Glacier Shrinkage in the Andes." For a general discussion, see World Wildlife Fund, "The Global Glacier Decline," www.panda.org/climate/glaciers.

42 Material in these paragraphs is from an interview with Lonnie Thompson, Byrd Polar Research Center, Ohio State University, 21 Sept. 2005; Lonnie Thompson et al., "Abrupt Tropical Climate Change: Past and Present," *Proceedings of the National Academy of Sciences* 103 (11 July 2006); Kevin Krajick, "Ice Man: Lonnie Thompson Scales the Peaks for Science," ibid.; Mark Bowen, "Thompson's Ice Corps," *Natural History,* Feb. 1998; Ned Rozell, "Where Science and High Adventure Meet," 23 Apr. 2004, www.sitnews.us/0404news/042304/042304_ak_science.html. Also see Mark Bowen, *Thin Ice: Unlocking the Secrets of Climate in the World's Highest Mountains* (New York: Henry Holt, 2005).

43 Lonnie Thompson et al., "Kilimanjaro Ice Core Records: Evidence of Holocene Climate Change in Tropical Africa," *Science* 298 (18 Oct. 2002). Scientists who disagree with Thompson's conclusion that global warming is primarily responsible for the loss of Kilimanjaro's glaciers believe that climate cycles and accelerated sublimation, not melting, are to blame. However, as one paper concludes, this "does not rule out that these processes may be linked to temperature variations" (Georg Kaser et al., "Modern Glacier Retreat on Kilimanjaro as Evidence of Climate Change: Observations and Facts," *International Journal of Climatology* 24 [2004]).

44 Vladimir Aizen et al., "Global Climate and Environmental Changes in Alpine Asia," in NSIDC, *Workshop on Assessing Global Glacier Recession;* Howard W. French, "A Melting Glacier in Tibet Serves as an Example and a Warning," *New York Times,* 9 Nov. 2004; Vladimir Aizen et al., "Glacier Change in Central and Northern Tien Shan during the Last 140 Years Based on Surface and Remote Sensing Data," *Annals of Glaciology* 43 (2006); Sandeep Chamling Rai, "An Overview of Glaciers, Glacier Retreat, and Subsequent Impacts in Nepal, India and China," WWF Nepal Program, Mar. 2005; "Glacier Study Reveals Chilling Prediction," People's Daily Online, 23 Sept. 2004, http://english.people.com.cn/200409/23/eng20040923_158036.html; Lonnie Thompson, email, 10 May 2006. For other Himalayan glaciers and the NASA project to monitor tens of thousands of glaciers, see http://asterweb.jpl.nasa.gov/content/03_data/05_Application_Examples/glacier/default.htm.

For one example of possible exceptions to the overall

retreat of glaciers, see Kenneth Hewitt, "The Karakoram Anomaly?" *Mountain Research and Development* 25 (Nov. 2005).

45 Barnett et al., "Potential Impacts of a Warming Climate"; "Qinghai-Tibet Glaciers Shrinking," *China Daily,* 5 Oct. 2004, www.chinadaily.com.cn/english/doc/2004–10/05/content _379858.htm.

46 Mark Meier, presentation at the 2002 American Association for the Advancement of Science meeting, Boston, MA, 16 Feb. 2002; Mark Meier and Mark Dyurgerov, "How Alaska Affects the World," *Science* 297 (19 Jul. 2002); Arendt et al., "Rapid Wastage"; Dyurgerov and Meier, "Glaciers and the Changing Earth System."

47 IPCC, "Climate Change 2007: The Physical Science Basis. Summary for Policymakers," 2 Feb. 2007, available at www .ipcc.ch; Sarah C. B. Raper and Roger J. Braithwaite, "Low Sea Level Rise Projections from Mountain Glaciers and Icecaps under Global Warming," *Nature* 439 (19 Jan. 2006); Eric Rignot, email, 9 Jan. 2006. See also Laury Miller and Bruce Douglas, "Mass and Volume Contributions to Twentieth-Century Global Sea Level Rise," *Nature* 428 (25 Mar. 2004).

48 Overpeck et al., "Paleoclimatic Evidence for Future Ice-Sheet Instability." Also see W. T. Pfeffer et al., "Kinematic Constraints on Glacier Contributions to 21st-Century Sea-Level Rise," *Science* 321 (5 Sept. 2008); "State of the Cryosphere" at http://nsidc.org/sotc.

CHAPTER 2

1 Interview with Bill Fraser, Palmer Station, Jan. 2000; updated via email Jan. 2006; Erik Stokstad, "Boom and Bust in a Polar Hot Zone," *Science* 315 (16 Mar. 2007).

2 See Raymond C. Smith et al., "Marine Ecosystem Sensitivity to Climate Change," *Bioscience* 49, no. 5 (1999); Steven Emslie et al., "Abandoned Penguin Colonies and Environmental Change in the Palmer Station Area, Anvers Is., Antarctic Peninsula," *Antarctic Science* 10, no. 3 (1998); J. P. Croxall et al., "Environmental Change and Antarctic Seabird Populations," *Science* 297 (30 Aug. 2002); response by D. G. Ainley et al., "Adélie Penguins and Environmental Change," *Science* 300 (18 Apr. 2003). Emslie comment is from "Penguin Study Offers Climate Clues," Environmental News Service, 8 Dec. 2003, at www.ens-newswire.com/ens/dec2003/2003–12–08–09 .asp#anchor7.

3 W. R. Fraser and E. E. Hoffman, "A Predator's Perspective on Causal Links between Climate Change, Physical Forcing and Ecosystem Response," *Marine Ecology Progress Series* 265 (31 Dec. 2003); Angus Atkinson et al., "Long-Term Decline in Krill Stock and Increase in Salps within the Southern Ocean," *Nature* 432 (4 Nov. 2004); Christophe Barbraud and Henri

Weimerskirch, "Emperor Penguins and Climate Change," *Nature* 411 (10 May 2001); Barbraud and Weimerskirch, "Antarctic Birds Breed Later in Response to Climate Change," *Proceedings of the National Academy of Sciences* 103 (18 April 2006). Also see J. P. Croxall et al., "Environmental Change and Antarctic Seabird Populations," in "Trouble in Polar Paradise" section of "News and Reviews," *Science* 297 (30 Aug. 2002). The Center for Biological Diversity cited global warming as "the most significant and pervasive threat" when it petitioned the USFWS in late 2006 to list twelve penguin species under the Endangered Species Act. See www.biologi caldiversity.org/swcbd/SPECIES/penguins/index.html.

4 Zimov et al., "Permafrost and the Global Carbon Budget"; K. M. Walter et al., "Methane Bubbling from Siberian Thaw Lakes as a Positive Feedback to Climate Warming," *Nature* 443 (7 Sept. 2006); Karen E. Frey and Laurence C. Smith, "Amplified Carbon Release from Vast West Siberian Peatlands by 2100," *Geophysical Research Letters* 32 (5 May 2005). Research background from Patrick Crill, email, July 2004; T. R. Christensen et al., "Thawing Sub-Arctic Permafrost: Effects on Vegetation and Methane Emissions," *Geophysical Research Letters* 31 (20 Feb. 2004). Other natural sources of methane are temperate peat bogs and wetlands, termites, and gas hydrates (also called clathrates, rigid water molecules surrounding molecules of natural gas), but about 60 percent of global methane emissions are from human-related activities (IPCC).

5 This and subsequent information is from interviews and field trips with Walt Oechel and his colleagues at Barrow Arctic Science Center, Aug. 2002. Temperature study is J. C. Comiso, "Warming Trends in Arctic from Clear Sky Satellite Observations," *Journal of Climatology* 16 (1 Nov. 2003). Also see confirmation from Canada in Philip Camill, "Permafrost Thaw Accelerates in Boreal Peatlands during Late-Twentieth-Century Climate Warming," *Climatic Change* 68 (January 2005).

6 Technical information in this section is from the group website, http://fs.sdsu.edu/gcrg; Walter Oechel et al., "Acclimation of Ecosystem CO_2 Exchange in the Alaskan Arctic in Response to Decadal Climate Warming," *Nature* 406 (31 Aug. 2000); "Ecosystem Interactions," in 2005 Annual Report, Barrow Environmental Observatory, BEO Subcommittee, Barrow Arctic Science Consortium, Jan. 2006; and Terry V. Callaghan et al., "Climate Change and UV-B Impacts on Arctic Tundra and Polar Desert Ecosystems: Key Findings and Extended Summaries," *Ambio* 33, no. 7 (Nov. 2004).

7 Terry Chapin, interview, Toolik, July 1999; Gus Shaver, interviews, Toolik , July 1999, and Barrow, Aug. 2002; Michelle C. Mack et al., "Ecosystem Carbon Storage in Arctic

An Environment Canada web page titled "Temperature and Precipitation in Historical Perspective" can be found at www.msc-smc.ec.gc.ca/ccrm/bulletin/national_e.cfm. See also W. A. Kurz et al., "Mountain Pine Beetle and Forest Carbon Feedback to Climate Change," *Nature* 452 (24 Apr. 2008); David D. Breshears et al., "Regional Vegetation Die-Off in Response to Global-Change-Type Drought," *Proceedings of the National Academy of Sciences* 102 (10 Oct. 2005).

19 "The 2000 Wildland-Urban Interface in the U.S.," at www.silvis.forest.wisc.edu/projects/US_WUI_2000.asp; National Climatic Data Center, *Climate of 2005—In Historical Perspective,* Annual Report, 13 Jan. 2006, available at www.ncdc.noaa.gov/oa/climate/research/2005/ann/ann05.html; Eric Kasischke, email, 20 Jan. 2006. Fire data at www.ncdc.noaa.gov/oa/climate/research/2007/fire07.html; A. L. Westerling et al., "Warming and Earlier Spring Increase Western U.S. Forest Wildfire Activity," *Science* 313 (18 Aug. 2006); A. J. Gillett et al., "Detecting the Effect of Climate Change on Canadian Forest Fires," *Geophysical Research Letters* 31 (29 Sept. 2004); U.N. Food and Agricultural Organization (FAO), "Global Forest Fire Assessment, 1990–2000," Forest Resources Assessment Working Paper 55, Rome, 2001; Global Fire Monitoring Network at www.gfmc.org; European Commission, "Forest Fires in Europe 2005," EUR 22312 EN, available at http://effis.jrc.it/documents/2006/ForestFiresInEurope2005.pdf.

20 Guido R. van der Werf et al., "Continental-Scale Partitioning of Fire Emissions during the 1997 to 2001 El Niño/La Niña Period," *Science* 303 (2 Jan. 2004); see also Susan E. Page et al., "The Amount of Carbon Released from Peat and Forest Fires in Indonesia during 1997," *Nature* 420 (7 Nov. 2002). Peatlands worldwide hold about 30 percent of soil carbon: see www.wetlands.org and www.peat-portal.net.

21 For forest carbon uptake, see Richard J. Norby et al., "Forest Response to Elevated CO_2 Is Conserved across a Broad Range of Productivity," *Proceedings of the National Academy of Sciences* 102 (13 Dec. 2005); Christian Korner et al., "Carbon Flux and Growth in Mature Deciduous Forest Trees Exposed to Elevated CO_2," *Science* 309 (26 Aug. 2005); James Heath et al., "Rising Atmospheric CO_2 Reduces Sequestration of Root-Derived Soil Carbon," *Science* 309 (9 Sept. 2005); and Eric Davidson and Ivan Janssens, "Temperature Sensitivity of Soil Carbon Decomposition and Feedbacks to Climate Change," *Nature* 440 (9 Mar. 2006). Other outcomes are reported at the Duke University FACE (free-air CO_2 enrichment) site, http://face.env.duke.edu/main.cfm. Current statistics at UN FAO, "State of the World's Forests," 2007, at www.fao.org/docrep/009/a0773e/a0773e00.htm; and

Gordon B. Bonan, "Forests and Climate Change: Forcings, Feedbacks, and the Climate Benefits of Forests," *Science* 320 (13 June 2008). Also see Sebastiaan Luyssaert et al., "Old-Growth Forests as Global Carbon Sinks," *Nature* 455 (11 Sept. 2008).

22 David W. Inouye et al., "Climate Change Is Affecting Altitudinal Migrants and Hibernating Species," *Proceedings of the National Academy of Sciences* 97 (15 Feb. 2000); Erik Beever, Peter F. Brussard, and Joel Berger, "Patterns of Apparent Extirpation among Isolated Populations of Pikas *(Ochotona princeps)* in the Great Basin," *Journal of Mammalogy* 84, no. 1 (Feb. 2003).

23 Clinton Epps et al., "Effects of Climate Change on Population Persistence of Desert-Dwelling Mountain Sheep in California," *Conservation Biology* 18 (Feb. 2004); James Brown et al., "Reorganization of an Arid Ecosystem in Response to Recent Climate Change," *Proceedings of the National Academy of Sciences* 94 (Sept. 1997).

24 Camille Parmesan et al., "Impacts of Extreme Weather and Climate on Terrestrial Biota," *Bulletin of the American Meteorological Society* 81, no. 3 (2000).

25 C. Drew Harvell et al., "Climate Warming and Disease Risks for Terrestrial and Marine Biota," *Science* 296 (21 June 2002).

26 Dean Roemmich and John McGowan, "Climatic Warming and the Decline of Zooplankton in the California Current," *Science* 267 (3 Mar. 1995); R. Veit et al., "Apex Marine Predator Declines Ninety Percent in Association with Changing Oceanic Climate," *Global Change Biology* 3 (Feb. 1997); J. McGowan et al., "The Biological Response to the 1977 Regime Shift in the California Current," *Deep Sea Research II* 50, 2003; Glen Martin, "Sea Life in Peril: Plankton Vanishing," *San Francisco Chronicle,* 12 July 2005, available at www.sfgate.com/cgi-bin/article.cgi?f=/c/a/2005/07/12/MNG8SDMMR01.DTL; Peterson, email, 15 Feb. 2006.

27 Jane Lubchenco, lecture presented at Wildlife Conservation Lecture Series, Portland, OR, 28 Feb. 2006; Francis Chan et al., "Emergence of Anoxia in the California Current Large Marine Ecosystem," *Science* 319 (15 Feb. 2008).

28 Raphael Sagarin, interview, 2001, with email update Jan. 2006; R. D. Sagarin et al., "Climate-Related Change in an Intertidal Community over Short and Long Time Scales," *Ecological Monographs* 69, no. 4 (1999); J. C. Barry et al., "Climate-Related, Long-Term Faunal Changes in a California Rocky Intertidal Community," *Science* 267 (15 Feb. 1995).

29 A. J. Southward, S. J. Hawkins, and M. T. Burrows, "Seventy Years' Observations of Changes in Distribution and Abundance of Zooplankton and Intertidal Organisms in the

Western English Channel in Relation to Rising Sea Temperature," *Journal of Thermal Biology* 20 (1995); also Gregory Beaugrand et al., "Reorganization of North Atlantic Marine Copepod Biodiversity and Climate," *Science* 296 (31 May 2002).

30 John Lanchbery, "Ecosystem Loss and Its Implications for Greenhouse Gas Concentration Stabilisation," Royal Society for the Protection of Birds, 2004, www.stabilisation2005.com/20_John_Lanchbery.pdf.

31 Camille Parmesan, "Climate and Species Range," *Nature* 382 (29 Aug. 1996).

32 C. Parmesan et al., "Poleward Shifts in Geographic Ranges of Butterfly Species Associated with Regional Warming," *Nature* 399 (10 June 1999).

33 Parmesan, interview, Montpellier, France, 5 July 2004; update by email, Jan. 2006; also see radio interview at www.earthsky.com/shows/edgeofdiscovery.php?date=20020521 and . . . =20040729.

34 Camille Parmesan and Gary Yohe, "A Globally Coherent Fingerprint of Climate Change Impacts across Natural Systems," *Nature* 421 (2 Jan. 2003); Camille Parmesan and Hector Galbraith, "Observed Impacts of Global Climate Change in the U.S.," Pew Center on Global Climate Change, Washington, DC, Nov. 2004. At about the same time, another group of scientists led by ornithologist Terry Root identified 147 published papers worldwide that directly correlate ten years or more of temperature change with a trait change in a plant or animal; see Root et al., "Fingerprints of Global Warming on Wild Animals and Plants," *Nature* 421 (2 Jan. 2003).

35 T. Root et al., "Human-Modified Temperatures Induce Species Changes: Joint Attribution," *Proceedings of the National Academy of Sciences* 102, May 24, 2005. Cynthia Rosenzweig et al., "Attributing Physical and Biological Impacts to Anthropomorphic Climate Change," *Nature* 453 (15 May 2008). See also "Notes from the Field: Climate Change Interview: Human-Modified Temperatures Induce Changes in Species," World Wildlife Fund, 19 May 2005, available at http://photos.panda.org/news_facts/newsroom/on_the_ground/index.cfm?uNewsID=20673.

36 IUCN, "Climate Change and Nature: Adapting for the Future," Gland, Switz., no date, www.iucn.org/themes/climate/docs/climateandnature.pdf; Williams, "Projected Distributions." The Stony Brook World Environmental Forum "Climate Change and the World's Protected Areas," held at the State University of New York at Stony Brook, 6–8 May 2005, was presided over by Richard Leakey. This section of the chapter is based on the presentations at that conference, along with Lovejoy and Hannah's book on biodiversity and climate change and Camille Parmesan's 2006 paper on ecology and evolutionary response.

37 William E. Bradshaw and Christina M. Holzapfel, "Evolutionary Response to Rapid Climate Change," *Science* 312 (9 Jun. 2006), and idem, "Genetic Shift in Photoperiodic Response Correlated with Global Warming," *Proceedings of the National Academy of Sciences* 98 (4 Dec. 2001); Jay Malcolm and Adam Markham, "Global Warming and Terrestrial Biodiversity Decline," report prepared for the World Wildlife Fund, 2000, available at http://assets.panda.org/downloads/speedkills_c6s8.pdf; see also Ronald Neilson et al., "Forecasting Regional to Global Plant Migration in Response to Climate Change," *Bioscience* 55 (9 Sept. 2005). Louis Pitelka and Plant Migration Workshop Group, "Plant Migration and Climate Change," *American Scientist* 85 (Sept.–Oct. 1997).

38 Erika Zavaleta et al., "Additive Effects of Simulated Climate Changes, Elevated CO_2, and Nitrogen Deposition on Grassland Diversity," *Proceedings of the National Academy of Sciences* 100 (24 June 2003); Marten Scheffer et al., "Catastrophic Shifts in Ecosystems," *Nature* 413 (11 Oct. 2001); Parmesan, "Ecological and Evolutionary Responses."

39 World Database on Protected Areas, www.unep-wcmc.org/wdpa; Secretariat of the Convention on Biological Diversity, "Interlinkages between Biological Diversity and Climate Change," CBD Technical Series 10, Montreal, 2003. The Human Footprint Project is at www.ciesin.columbia.edu/wild_areas; McNeeley quote is from the Stony Brook Forum 2005.

40 Pitelka, "Plant Migration and Climate Change"; for migratory species, see UNEP and Convention on the Conservation of Migratory Species, "Migratory Species and Climate Change: Impacts of a Changing Environment on Wild Animals" (2006), available at www.cms.int/publications/pdf/CMS_CimateChange.pdf; Leakey quote is from the Stony Brook Forum, 2005.

41 List of threatened parks is from presentations at the Stony Brook Forum and news reports. See also "Hidden Garden of Eden Wilts in Soaring Heat," *New Scientist,* 11 Mar. 2006.

42 Species numbers are from Nalini Nadkarni and Nathaniel Wheelwright, eds., *Monteverde* (New York: Oxford University Press, 2000), 143–44.

CHAPTER 4

1 In the twentieth century, the Earth's average surface air temperature went through two warming periods with a slight cooling in midcentury, resulting in a 1.1°F (0.6°C) rise. However, when scientists look over data from 1905 to 2005, they see that temperature rose 1.4°F (0.8°C). Thus, the rate of change is increasing. If the current rate continues over the course of the present century, this will mean a hundred-year increase of 3.6°F (2°C). See James Hansen et al., "GISS Surface Temperature Analysis: Global Temperature Trends—2005 Summation," http://data.giss.nasa.gov/gistemp/2005.

A distinct signature of greenhouse warming is that CO_2 warms the troposphere but cools the stratosphere and above. Ozone depletion cools the stratosphere as well. See Liying Qian et al., "Calculated and Observed Climate Change in the Thermosphere, and a Prediction for Solar Cycle 24," *Geophysical Research Letters* 33 (6 Dec. 2006); J. Lastovicka et al., "Global Change in the Upper Atmosphere," *Science* 314 (24 Nov. 2006); and ESPERE, "Upper Atmosphere," www.atmosphere.mpg.de/enid/20c.html. Molina quote is from Stony Brook World Environmental Forum, State University of New York at Stony Brook, 7 May 2005.

2 S. Levitus, "Warming of the World Ocean, 1955–2003," *Geophysical Research Letters* 32 (22 Jan. 2005); Tim P. Barnett et al., "Penetration of Human-Induced Warming into the World's Oceans," *Science* 309 (8 July 2005). The average temperature increase between 1955 and 2003 of the oceans down to 9,800 feet is less than 0.1°F, yet it occurred around the world and indicates a great amount of heat—strong confirmation of the energy added from human greenhouse gases. C. L. Sabine et al., "The Oceanic Sink for Anthropogenic CO_2," *Science* 305 (16 July 2004); Corinne Le Quéré et al., "Saturation of the Southern Ocean CO_2 Sink Due to Recent Climate Change," *Science* 316 (22 June 2007); sea level rise statistics from Kevin Trenberth, National Center for Atmospheric Research, 21 Oct. 2004, quoted at www.ucar.edu/news/record. See also "Indicator: Sea Level (353)," in *National Report on the Environment,* Environmental Protection Agency, Washington, DC, 2006, at www.epa.gov/ncea/ROEIndicators/pdfs/SEA LEVEL_FINAL.pdf. For effects on human life and cities throughout this chapter, see IPCC, "Climate Change 2007: Impacts, Adaptation, and Vulnerability. Summary for Policymakers," 6 Apr. 2007, available at www.ipcc.ch.

3 Tuvalu's carbon emissions per capita are estimated at half a ton, but Tuvalu is not listed on the UN statistics because the total amount is so slight compared to most other nations. Quotations in this section are from interviews conducted 8–13 Feb. 2005 in Funafuti; some facts are from www.indexmundi.com/tuvalu/index.html#demographics.

4 Lower estimate of sea level rise is from John R. Hunter, "A Note on Relative Sea Level Change at Funafuti, Tuvalu," Antarctic Cooperative Research Centre, Hobart, Tas., Austr., 12 Aug. 2002, available at http://staff.acecrc.org.au/~johnhunter/tuvalu.pdf; higher estimate is from AusAID, "Sea Level and Climate: Their Present State, Tuvalu," Pacific Country Report, Flinders, Vic., Austr., June 2003. For 2006 tides, see www.tuvaluislands.com/news/archives/2006/

2006–03–04.htm; and Samir S. Patel, "A Sinking Feeling," *Nature* 440 (6 Apr. 2006).

5 Alexander Berzon, "Tuvalu Is Drowning," www.salon.com/news/feature/2006/03/31/tuvalu/index_np.html.

6 J. Barnett and W. N. Adger, "Climate Dangers and Atoll Countries," *Climatic Change* 61, no. 3 (2003); "Island Communities Are Lost before the Sea Level Rises," Friends of the Earth, Australia, 2003, at www.foei.org/publications/pdfs/island.pdf; Tuvaluan governor-general's comment is from "Citizen's Guide to Climate Refugees," Friends of the Earth/Australia, 2005, www.foe.org.au/download/CitizensGuide.pdf.

7 Don Hinrichsen, "Ocean Planet in Decline," 12 Oct. 2004, www.peopleandplanet.net/doc.php?id=429&section=6; Bruce C. Douglas and W. Richard Peltier, "The Puzzle of Global Sea-Level Rise," *Physics Today* 55 (Mar. 2002); Jeff Hecht, "Disappearing Deltas Could Spell Disaster," *New Scientist,* 18 Feb. 2006; Asian Delta Project, at http://unit.aist.go.jp/igg/rg/cug-rg/ADP/ADP_E/a_igcp475_en.html; "Reducing Risks to Cities from Disasters and Climate Change, Apr. 2007, available at http://eau.sagepub.com.

8 Philip Gain, *Bangladesh Environment: Facing the 21st Century* (Dhaka: Society for Environment and Human Development, 2002).

9 Ibid. See also Saleemul Huq, "Climate Change and Bangladesh," *Science* 294 (23 Nov. 2001).

10 Notes from meeting with NAOMI retired Bangladesh scientist group, Dhaka, 23 June 2005; S. Adrawala et al., "Development and Climate Change in Bangladesh: Focus on Coastal Flooding and the Sundarbans," OECD, Paris, 2003, www.oecd.org/dataoecd/46/55/21055658.pdf.

11 Interviews with residents of Bhola Island, 24–26 June 2005.

12 Anwar Ali, "Vulnerability of Bangladesh Coastal Region to Climate Change with Adaptation Options," Bangladesh Space Research and Remote Sensing Organization (SPARRSO), Dhaka, Bangladesh, no date (after 2000?); Intergovernmental Panel on Climate Change, *Climate Change 2001: Summary for Policymakers* (Cambridge: Cambridge University Press, 2001).

13 History and details of Dutch protections against the sea are at www.deltawerken.com (English pages); economic figures and additional history from Jan R. Hoogland, "Safety in the Netherlands," Statement to the United States Congress, 20 Oct. 2005, at www.house.gov/transportation/water/10–20–05/hoogland.pdf; interview with Dutch development official Ton von Oschee, The Hague, Aug. 2004.

14 Joseph A. Harriss, "Turning the Tide," *Smithsonian,* Sept. 2002; "Sea Level Rise and Coastal Disasters," summary

of a forum, Natural Disasters Roundtable, National Academy of Sciences, Washington, DC, 25 Oct. 2001, at www .nap.edu/catalog/php?record_id=10590#toc; "Defence from High Waters," at www.salve.it/uk/soluzioni/problemi/ P-eccezionaliA.htm.

15 Sir David King, science adviser to the British government, on NPR's "Talk of the Nation," 13 May 2005; Stefan Rahmstorf, "A Semi-Empirical Approach to Projecting Future Sea-Level Rise," *Science Express* (14 Dec. 2006), at www .sciencemag.org/cgi/rapidpdf/1135456v1. According to British government reports and Parliamentary testimony, the Thames Barrier is raised for other reasons as well, such as controlling river flow, but King's figure is only the number resulting from tidal flooding. For an interactive map showing the land that will be inundated as sea levels rise, see http://geongrid.geo .arizona.edu/arcims/website/slrworld/viewer.htm.

16 "Scientists' Fears Come True as Hurricane Floods New Orleans," *Science* 309 (9 Sept. 2005); Lester Brown, "Global Warming Forcing U.S. Coastal Population to Move Inland," Earth Policy Institute, 16 Aug. 2006, www.earth-policy .org/Updates/2006/Update57.htm; Stephen Leatherman, "10 Most Hurricane Vulnerable Areas," Florida International University, available at www.ihc.fiu.edu/media/docs/ 10_Most_Hurricane_Vulnerable_Areas.pdf; "Climate Change Futures: Health, Ecological, and Economic Dimensions," Center for Health and the Global Environment, Harvard University, Cambridge, MA, Nov. 2005.

17 Kevin Trenberth, "Uncertainty in Hurricanes and Global Warming," *Science* 308 (17 June 2005); and email, 5 Feb. 2006.

18 NCDC quote is from http://lwf.ncdc.noaa.gov/oa/climate/ research/2005/katrina.html; Kerry Emanuel, "Increasing Destructiveness of Tropical Cyclones over the Past 30 Years," *Nature* 436 (4 Aug. 2005); P. J. Webster et al., "Changes in Tropical Cyclone Number, Duration, and Intensity in a Warming Environment," *Science* 309 (16 Sept. 2005); Trenberth, email, 5 Feb. 2006; K. Emanuel et al., "Hurricanes and Global Warming: Results from Downscaling IPCC AR4 Simulations." *Bulletin of the American Meteorological Society* 89 (March 2008). Other studies using various methods diverge widely in correlation with climate warming; see next note and discussions on www.realclimate.org.

19 Gray's hurricane forecast is at http://hurricane.atmos .colostate.edu/Forecasts/2006/aug2006/; Patrick Michaels et al., "Sea-Surface Temperatures and Tropical Cyclones in the Atlantic Basin," *Geophysical Research Letters* 33 (10 May 2006); Amato T. Evan et al., "New Evidence for a Relationship between Atlantic Tropical Cyclone Activity and African Dust Outbreaks," *Geophysical Research Letters* 33 (10 Oct. 2006); another dissenting study is Philip Klotzbach, "Trends in

Global Tropical Cyclone Activity over the Past Twenty Years," *Geophysical Research Letters* 33 (20 May 2006). Further analysis supporting the role of sea temperatures and discounting regional oscillations is in M. E. Mann and K. A. Emanuel, "Atlantic Hurricane Trends Linked to Climate Change," *Eos* 87 (13 June 2006). Also see Judith Curry et al., "Mixing Politics and Science in Testing the Hypothesis That Greenhouse Warming Is Causing a Global Increase in Hurricane Intensity," *Bulletin of the American Meteorological Society,* Aug. 2006. For hurricane data, see www.ncdc.noaa.gov/oa/climate/ research/2006/hurricanes06.html and http://weather.unisys .com/hurricane/index.html. The WMO "Statement on Tropical Cyclones and Climate Change," Nov. 2006, is at www .wmo.ch/web/arep/arep-home.html; see also James Elsner et al., "The Increasing Intensity of the Strongest Tropical Storms," *Nature* 455 (4 Sept. 2008); P. W. Thorne, "The Answer Is Blowing in the Wind," *Nature Geoscience* 6 (Jun. 2008).

20 For an argument that damage from storms is a more meaningful measure than storm intensity, see R. A. C. Pielke Jr. et al., "Hurricanes and Global Warming," *Bulletin of the American Meteorological Society* 86 (Dec. 2005); population figures from NOAA, "National Overview of Population Trends along the Coastal United States," www.oceanservice .noaa.gov/programs/mb/pdfs/2_national_overview.pdf.

21 "Global Climate Change . . . Here and Now," conference of the North Carolina Coastal Federation, 1 Oct. 2004, Morehead City, NC; Jan DeBlieu, interview and tour of Cape Hatteras, 30 Sept. 2004.

22 K. Zhang, B. C. Douglas, and S. P. Leatherman, "Global Warming and Coastal Erosion," *Climatic Change* 64 (May 2004); DeBlieu interview. For more information on beaches, consult Wallace Kaufman and Orrin H. Pilkey Jr., *The Beaches Are Moving: The Drowning of America's Shoreline* (Chapel Hill, NC: Duke University Press, 1983).

23 Hector Galbraith et al., "Global Climate Change and Sea Level Rise: Potential Losses of Intertidal Habitat for Migrating Shorebirds," *Waterbirds* 25 (June 2002). The loss of feeding habitat is part of a dual threat to red knots, whose 10,000-mile (16,000 km) migration to Alaska from South America is one of the longest of any bird. Their prime refueling food on the Delaware Bay is eggs of horseshoe crabs, whose population is crashing. Numbers of red knots in recent migrations were the lowest ever recorded, about 10 percent of previous numbers.

24 Baltimore tide gauge data can be seen at www.ngs.noaa .gov/GRD/GPS/Projects/CB/SEALEVEL/sealevel.html; Alan Elsner, "Wildlife Preserve Shows Effect of Global Warming," Reuters, 13 Mar. 2001, available at www.planetark.org/daily newsstory.cfm?newsid=10092; Michael Kearney et al., "Landsat Imagery Shows Decline in Coastal Marshes in Chesapeake and Delaware Bays," *Eos* 83 (16 Apr. 2002).

25 "Climate of 2005: July in Historical Perspective," www .ncdc.noaa.gov/oa/climate/research/2005/jul/jul05.html; "Billion Dollar U.S. Weather Disasters, 1980–2005," www .ncdc.noaa.gov/oa/reports/billionz.html#narrative; Edward R. Cook et al., "Long-Term Aridity Changes in the Western United States," *Science* 306 (5 Nov. 2004). For more technical views on weather trends, see "On Record-Breaking Events," www.realclimate.org/index.php?p=175. "Extreme" weather means events that are extremely rare, set new records across a long time span, or are repeated occurrences of weather maximums. In the United States, for example, "very heavy" rainfall means a deluge of more than 3 to 4 inches (75–100 mm) in a day—something that occurs only 0.3 percent of the time. See "Extreme Weather and Climate Events and Public Health Responses," report on a WHO meeting held in Bratislava, Slovakia, 9–10 Feb. 2004, available at www.euro.who .int/eprise/main/WHO/Progs/GCH/Topics/20040116_1? language=; Pavel Ya. Groisman et al., "Contemporary Changes of the Hydrological Cycle over the Contiguous United States: Trends Derived from In Situ Observations," *Journal of Hydrometeorology* 5 (2004).

26 "Billion-Dollar U.S. Weather Disasters."

27 "Topics Geo: Annual Review—Natural Catastrophes 2005," Munich Re, available at www.munichre.com (search for publications and natural catastrophes on the internal search engine); "Climate Change Futures"; Swiss Re quote is at www.swissre.com/internet/pwswpspr.nsf/fmBookMark FrameSet?ReadForm&BM=../vwAllbyIDKeyLu/bmer-6 mak8s?OpenDocument; "Insurance Companies Staggering under Global Warming Damages," available at www.organic consumers.org/Politics/insurance091505.cfm; Joel Garreau, "A Dream Blown Away," *Washington Post* (2 Dec. 2006), available at www.washingtonpost.com/wp-dyn/content/ article/2006/12/01/AR2006120101759.html.

28 John Magnuson et al., "Historical Trends in Lake and River Ice Cover in the Northern Hemisphere," *Science* 289 (8 Sept. 2000); Raphael Sagarin and Fiorenza Micheli, "Climate Change in Nontraditional Data Sets," *Science* 294 (26 Oct. 2001); Pavel Ya. Groisman et al., "Changes of Snow Cover, Temperature, and Radiative Heat Balance over the Northern Hemisphere," *Journal of Climate* 7, no. 11 (1994); Michael Dettinger, "Changes in Streamflow Timing in the Western United States in Recent Decades," USGS Fact Sheet 2005– 3018, Mar. 2005; Satish Kumar Regonda et al., "Seasonal Cycle Shifts in Hydroclimatology over the Western United States," *Journal of Climate* 18 (Jan. 2005). Other weather statistics here and below are from Easterling et al., "Climate Extremes: Ob-

servations, Modeling, and Impacts," *Science* 289 (22 Sept. 2000).

29 Thomas R. Karl and Kevin E. Trenberth, "Modern Global Climate Change," *Science* 302 (5 Dec. 2003); rainfall figures from Pavel Ya. Groisman et al., "Trends in Intense Precipitation in the Climate Record," *Journal of Climate* 18, no. 9 (2005); B. N. Goswami et al., "Increasing Trend of Extreme Rain Events over India in a Warming Environment," *Science* 314 (1 Dec. 2006); "Climate Change, Rivers, and Rainfall," Hadley Centre, Met Office, Exeter, Eng., Dec. 2005, available at www.metoffice.com/research/hadleycentre/pubs/brochures/; "Climate Change and River Flooding in Europe," European Environment Agency, Copenhagen, 2005, available at http:// reports.eea.europa.eu/briefing_2005_1/en.

30 Drought statistics from A. Dai, K. E. Trenberth, and T. Qian, "Global Dataset of Palmer Drought Severity Index for 1870–2002: Relationship with Soil Moisture and Effects of Surface Warming," *Journal of Hydrometeorology* 5, no. 6 (2004), a paper that is summarized in "Drought's Growing Reach: NCAR Study Points to Global Warming as Key Factor," www.ucar.edu/news/releases/2005/drought_research.shtml; Gaia Vince, "Dust Storms on the Rise Globally," *New Scientist*, 20 Aug. 2004, www.newscientist.com/article.ns?id=dn6306; Wolfgang Buermann et al., "The Changing Carbon Cycle at Mauna Loa Observatory," *Proceedings of the National Academy of Sciences* 104 (13 Mar. 2007).

31 David Travis et al., "Climatology: Contrails Reduce Daily Temperature Range," *Nature* 418 (8 Aug. 2002); Nicola Stuber et al., "The Importance of the Diurnal and Annual Cycle of Air Traffic for Contrail Radiative Forcing," *Nature* 441 (15 June 2006).

32 Global dimming is a difficult issue because the mechanisms of cloud formation and effects of aerosols are poorly understood. See "Could Reducing Global Dimming Mean a Hotter, Dryer World?" www.ldeo.columbia.edu/news/ 2006/04_14_06.htm; and Beate Liepert's explanation at www .ldeo.columbia.edu/~liepert/research/globalDimming.html. Also see global dimming discussion (somewhat technical) at www.realclimate.org/index.php?p=105; and Meinrat O. Andreae et al., "Strong Present Day Aerosol Cooling Implies a Hot Future," *Nature* 435 (30 June 2005).

33 "Climate Change and Human Health—Risks and Responses," WHO, Geneva, 2003, available at www.who.int/ globalchange/publications/cchhsummary/en; *Montreal Star*, 9 Dec. 2005. Sir David King, British science adviser, stated a figure of 160,000 deaths at the American Association for the Advancement of Science annual meeting, Seattle, Jan. 2004. See also "Extreme Weather Events and Climate Instability: Implications for Human Health," Center for

Health and the Human Environment, Harvard University, Sept. 2004.

34 Paul Reiter, "Climate Change and Highland Malaria in the Tropics," paper delivered at conference titled "Avoiding Dangerous Climate Change," Met Office and DEFRA, Exeter, Eng., 1–3 Feb. 2005 (reports and abstracts available at www.stabilisation2005.com); Paul R. Epstein et al., "Biological and Physical Signs of Climate Change: Focus on Mosquito-Borne Diseases," *Bulletin of the American Meteorological Society* 79 (Mar. 1998); M. Pascual, "Malaria Resurgence in the East African Highlands: Temperature Trends Revisited," *Proceedings of the National Academy of Sciences* 103 (11 Apr. 2006); http://westnilemaps.usgs.gov. Additional West Nile information can be found in "Impacts of a Warming Arctic: Highlights" (brochure), www.amap.no/acia/Highlights.pdf; and Centers for Disease Control pages at www.cdc.gov/ncidod/dvbid/westnile/index.htm. Jonathan A. Patz et al., "Impact of Regional Climate Change on Human Health," *Nature* 438 (17 Nov. 2005). See also Paul Epstein, "Is Global Warming Harmful to Health?" *Scientific American,* Aug. 2000.

35 Peter Gleick, "Global Freshwater Resources: Soft-Path Solutions for the Twenty-first Century," *Science* 302 (28 Nov. 2003); "Climate Change, Rivers, and Rainfall"; "Public Health Impacts," Program of the Health Effects of Global Environmental Change, Johns Hopkins University, www.jhu.edu/~climate/health.html; *Ecosystems and Human Well-Being: Synthesis,* Millennium Ecosystem Assessment (Washington, DC: Island Press, 2005), also available at www.maweb.org/en/products.aspx; Patz et al., "Impact of Regional Climate Change."

36 Andrew Simms and Hannah Reid, "Africa—Up in Smoke?" and "Africa—Up in Smoke 2," Working Group on Climate Change and Development, New Economics Foundation, London, June 2005/2006. Flood information from Kenya travel and interviews with Peter Smerdon, World Food Agency, Nairobi, 9–20 Nov. 2006. Floods in equatorial east Africa also occurred in 1998, attributed to El Niño; those of late 2006 appear to have the same meteorological signature but are being called the worst in fifty years. See www.cpc.noaa.gov/products/analysis_monitoring/enso_advisory.

37 Martin Hoerling et al., "Detection and Attribution of Twentieth-Century Northern and Southern African Rainfall Change," *Journal of Climate* 19 (Aug. 2006); Martin Hoerling, email and telephone, 15 and 16 June 2006; J. W. Hurrell and M. P. Hoerling, "The Great Twentieth-Century Drying of Africa," *Eos,* Transactions of the American Geophysical Union, 86, no. 18, Joint Assembly Supplement, Abstract A22A-02, May 2005; and Jian Lu and Thomas Delworth, "Oceanic

Forcing of the Late-Twentieth-Century Sahel Drought," *Geophysical Research Letters* 32 (19 Nov. 2005). I also consulted digests of papers about the Sahel by Aiguo Dai of NCAR, such as "The Recent Sahel Drought Is Real," *International Journal of Climatology* 24 (2004); A. Giannini et al., "Oceanic Forcing of Sahel Rainfall on Interannual to Interdecadal Time Scales," *Science* 302 (7 Nov 2003).

38 A. B. Pezza and I. Simmonds, "The First South Atlantic Hurricane: Unprecedented Blocking, Low Shear, and Climate Change," *Geophysical Research Letters* 32 (12 Aug. 2005).

39 Amazon Environmental Research Institute, "Amazon Basin Experiencing Extreme Drought," 19 Oct. 2005. www.forests.org/articles/reader.asp?linkid=47478; Paul Lefebvre, Woods Hole Research Center, email, 24 Mar. 2006; "Was the Record Amazon Drought Caused by Warm Seas?" www.realclimate.org/index.php?p=230; Gabriel A. Vecchi et al., "Weakening of Tropical Pacific Atmospheric Circulation Due to Anthropogenic Forcing," *Nature* 441 (4 May 2006); Marcos D. Oyama and Carlos A. Nobre, "A New Climate-Vegetation Equilibrium State for Tropical South America," *Geophysical Research Letters* 30 (5 Dec. 2003); Daniel Nepstad et al., "Amazon Drought and Its Implications for Forest Flammability and Tree Growth: A Basin-Wide Analysis," *Global Change Biology* 10 (May 2004); L. R. Hutyra et al., "Climatic Variability and Vegetation Vulnerability in Amazonia," *Geophysical Research Letters* 32 (24 Dec. 2005); F. J. F. Chagnon and R. L. Bras, "Contemporary Climate Change in the Amazon," *Geophysical Research Letters* 32 (9 July 2005); Jim Giles, "The Outlook for Amazonia Is Dry," *Nature* (16 Aug. 2006), www.nature.com/news/2006/060814/full/442726c.html; deforestation rate from www.nasa.gov/centers/goddard/news/topstory/2005/amazon_deforest.html. A useful review of climate effects over Amazonia is " Climate Change Impacts in the Amazon: Review of Scientific Literature," World Wildlife Fund, Gland, Switz., 2005, at http://assets.panda.org/downloads/amazon_cc_impacts_lit_review_final_2.pdf. For more on global warming in Latin America, see "Up in Smoke? Latin America and the Caribbean," Working Group on Climate Change and Development, at www.neweconomics.org/gen/uploads/15erpvfzxbbipu552pn001f128082006213002.pdf.

40 Health and mortality information from WHO, "Extreme Weather and Climate Events"; UN Environment Programme, "Impacts of Summer 2003 Heat Wave in Europe," GRID-Europe, Mar. 2004, www.grid.unep.ch/product/publication/download/ew_heat_wave.en.pdf; U.S. deaths from "Natural Hazard Statistics: Weather Fatalities," at www.nws.noaa.gov/om/hazstats.shtml; Gerald Meehl and Claudia Tebaldi, "More Intense, More Frequent, and Longer-Lasting Heat Waves in the Twenty-first Century," *Science* 305 (13 Aug.

2004); the tie to gases is from Peter Stott, D. A. Stone, and M. R. Allen, "Human Contribution to the European Heat Wave of 2003," *Nature* 432 (2 Dec. 2004); U.K. Met Office press release of 1 Dec. 2004; Christoph Schar et al., "The Role of Increasing Temperature Variability in European Summer Heat Waves," *Nature* 427 (22 Jan. 2004); Phillippe Ciais et al., "Europe-Wide Reduction in Primary Production Caused by the Heat and Drought in 2003," *Nature* 437 (22 Sept. 2005). An overview of heat wave occurrence and research can be found in Juliet Eilperin, "More Frequent Heat Waves Linked to Global Warming," *Washington Post,* 4 Aug. 2006. See also Paul R. Epstein and Christine Rogers, "Inside the Greenhouse: The Impacts of CO_2 and Climate Change on Public Health in the Inner City," Center for Health and the Global Environment, Harvard Medical School, Apr. 2004, at http://chge.med.harvard.edu/publications/documents/green.pdf.

41 Food quote is from John R.Christy at "Global Climate Change," 11th Annual Symposium of the Stegner Center for Land, Resources, and the Environment, University of Utah, Salt Lake City, 4 Mar. 2006; Fred Pearce, "Climate Change Warning over Food Production," *New Scientist,* 26 Apr. 2005, www.newscientist.com/channel/earth/dn7310-climate-change-warning-over-food-production.html; Tina Hesman, "Greenhouse Gassed," *Science News* 157 (25 Mar. 2000); "CO_2 Fertilization: Boon or Bust," panel discussion at annual meeting of the American Association for the Advancement of Science (AAAS), Seattle, 16 Feb. 2004; Peter Reich et al., "Nitrogen Limitation Constrains Sustainability of Ecosystem Response to CO_2," *Nature* 440 (13 Apr. 2006); Glenn Scherer, "The Food, the Bad, and the Ugly," *Grist,* 12 July 2005, www.grist.org/news/maindish/2005/07/12/scherer-plantchem; Jacqueline E. Mohan et al., "Biomass and Toxicity Responses of Poison Ivy *(Toxicodendron radicans)* to Elevated Atmospheric CO_2," *Proceedings of the National Academy of Sciences* 103 (13 June 2006).

42 Long's research findings are included in "Food Crops in a Changing Climate: Report of a Royal Society Discussion Meeting Held April 2005," policy document 10/05, Royal Society, June 2005, www.royalsoc.ac.uk/displaypagedoc.asp?id=13105; Jim Giles, "Photosynthesis Threatened by Rising Air Pollution," *Nature* (4 May 2005); David B. Lobell and Gregory P. Asner, "Climate and Management Contributions to Recent Trends in U.S. Agricultural Yields," *Science* 299 (14 Feb. 2003); David Lobell and Christopher Field, "Global Scale Climate-Crop Yield Relationships and the Impacts of Recent Warming," *Environmental Research Letters* 2 (Mar. 2007); Shaobing Peng et al., "Rice Yields Decline with Higher Night Temperature from Global Warming," *Proceedings of the National Academy of Sciences* 101 (6 July 2004). Regarding other

agricultural products, see, e.g., M. A. White et al., "Extreme Heat Reduces and Shifts United States Premium Wine Production in the Twenty-first Century," *Proceedings of the National Academy of Sciences* 103 (25 July 2006).

43 Paulay quote from Ninth International Coral Reef Symposium (ICRS-9), Bali, Indonesia, Oct. 2000.

44 J. M. Pandolfi et al., "Are U.S. Coral Reefs on the Slippery Slope to Slime?" *Science* 307 (18 Mar. 2005); Dirk Bryant et al., "Reefs at Risk," World Resources Institute, Washington, DC, 1998, available at www.wri.org/biodiv/pubs_description.cfm?pid=2901.

45 Ove Hoegh-Guldberg and Hans Hoegh-Guldberg, "Implications of Climate Change for Australia's Great Barrier Reef," WWF-Australia, 2004, updated Jan. 2006, available at www.wwf.org.au/news/n65.

46 Varon quote from ICRS-9, 2000; O. Hoegh-Guldberg, "Climate Change, Coral Bleaching, and the Future of the World's Coral Reefs," *Marine and Freshwater Research* 50 (1999), available at www.reef.edu.au/OHG/res-pic/HG%20papers/Hoegh-Guldberg%201999.pdf.

47 T. P. Hughes et al., "Climate Change, Human Impacts, and the Resilience of Coral Reefs," *Science* 310 (15 Aug. 2003); Hoegh-Guldberg, "Climate Change, Coral Bleaching, and the Future of the World's Coral Reefs"; Hoegh-Guldberg and Hoegh-Guldberg, "Implications of Climate Change for Australia's Great Barrier Reef"; Clive Wilkinson, "State of the Coral Reefs of the World: 2002," Australian Institute of Marine Science, Townsville, Qsld., 2004; J. M. Lough, "Sea Surface Temperatures on the Great Barrier Reef: A Contribution to the Study of Coral Bleaching," GBRMPA Research Publication No. 57, Townsville, Qsld., 1999. Also see Julia Whitty, "Shoals of Time," *Harper's,* Jan. 2001.

48 R. Aronson et al., "Coral Bleach-out in Belize," *Nature* 405 (4 May 2000), as well as remarks at AAAS meeting, Seattle, 14 Feb. 2004; Patrick Joy, "NOAA Scientist: 2005 Coral Bleaching Unprecedented," *Virgin Island Daily News,* 25 Jan. 2006; A. Bruckner, "Response to the 2005 Western Atlantic Coral Bleaching Event," NOAA Fisheries (draft), 4 Nov. 2005, http://coralreefwatch.noaa.gov/caribbean2005/docs/CRCP_draft_response.pdf; "Coral Bleaching and Disease Deliver 'One-Two Punch' to Coral Reefs in the U.S. Virgin Islands," National Park Service, Oct. 2006, at www.coralreef.gov/library/pdf/DOI%20bleaching%2011-06.pdf; Mark Drew, Nature Conservancy, email, 10 Apr. 2006. For more information on the 2005 bleaching, see http://coralreefwatch.noaa.gov/caribbean2005.

49 Hoegh-Guldberg, phone and email interviews, Feb. and Nov. 2005 and May 2006; Nicholas Graham et al., "Dynamic Fragility of Oceanic Coral Reef Ecosystems," *Pro-*

ceedings of the National Academy of Sciences 103 (30 May 2006).

50 Royal Society, "Ocean Acidification due to Increasing Atmospheric Carbon Dioxide," policy document 12/05, June 2005, available at www.royalsoc.ac.uk/document.asp ?id=3249. Figures on take-up and effect of CO_2 in oceans are in Christopher Sabine et al., "The Oceanic Sink for Anthropogenic CO_2," and Richard Feely et al., "Impact of Anthropogenic CO_2 on the $CaCO_3$ System in the Oceans," both in *Science* 305 (16 July 2004). Remarks by Joan Kleypas at ICRS-9, 2000, and AAAS meeting, Seattle, 2004, with an email update May 2006; O. Hoegh-Guldberg et al., "Coral Reefs under Rapid Climate Change and Ocean Acidification," *Science* 318 (14 Dec. 2007); Kent E. Carpenter et al., "One-Third of Reef-Building Corals Face Elevated Extinction Risk from Climate Change and Local Impacts," *Science Express* (10 Jul. 2008); Richard Feely, NOAA, seminar at the American Association for the Advancement of Science annual meeting, San Francisco, 16 Feb. 2007. According to the geologic record, Kleypas notes, corals have in the past lived under conditions of high CO_2, but they did not create reefs.

51 Royal Society, "Ocean Acidification" (2005); Michael Behrenfeld et al., "Climate-Driven Trends in Contemporary Ocean Productivity," *Nature* 444 (7 Dec. 2006).

52 Ray Berkelmans and Madeleine J. H. van Oppen, "The Role of Zooxanthellae in the Thermal Tolerance of Corals," *Proceedings of the Royal Society B* 273 (22 Sept. 2006); Rob Rowan, "Coral Bleaching: Thermal Adaptation in Reef Coral Symbionts," *Nature* 430 (12 Aug. 2004); Andrew C. Baker et al., "Coral Reefs: Corals' Adaptive Response to Climate Change," *Nature* 430 (12 Aug. 2004); John McWilliams et al., "Temperature-Induced Coral Bleaching in the Caribbean," *Ecology* 86 (8 Aug. 2005); Wilkinson, "State of the Coral Reefs of the World"; Ove Hoegh-Guldberg, telephone interview, Brisbane, Qsld., 28 Feb. 2005, and email, 8 Jan. 2006.

53 Hoegh-Guldberg and Hoegh-Guldberg, "Implications of Climate Change for Australia's Great Barrier Reef"; "Climate Change—Facts," WWF-Australia, available at wwf.org.au/ ourwork/climatechange/whatis. Also see Climate Action Network Australia website, www.cana.net.au; and www.bom .gov.au/Announcements/media_releases/climate/change.

54 "Storms Strike Country's East," *Queensland Times,* 4 Feb. 2005; general storm coverage in the *Australian,* 4 Feb. 2005; Lisa Yallamas, "Southeast Works Up a Thirst," *Courier-Mail,* 22 Jan. 2008; "Australia Records Hottest Spring," *The Age,* 15 Dec. 2006, available at www.theage.com.au/news/national/ australia-records-hottest-spring/2006/12/15/1165685857289 .html; Australian Bureau of Meteorology, "Drought State-

ment," 3 July 2008, at www.bom.gov.au/climate/drought/ drought.shtml; Alex Koutts, "Fires Likely to Kill Off Many of Australia's Native Species," *New Zealand Herald,* 20 Dec. 2006, available at www.nzherald.co.nz/section/2/story.cfm?c_ id=2&objectid=10416153.

55 Energy figures from Energy Information Administration, http://tonto.eia.doe.gov/country/country_energy_data.cfm ?fips=CH.

56 Money and population figures from Chinese sources via Li Moxuan, Greenpeace/China, Beijing, Jan. 2005; and Population Reference Bureau at www.prb.org; Greenpeace untitled report Mar. 2004 (via email), based on information from Qingyuan Hydrological Station and Guangdong Meteorological Bureau, www.greenpeace.org/china/en/news/20050603 _drought_gd_html; "China Farmland Hit by Drought up 21 Percent over 2005—Report," Reuters, 11 Aug. 2006, at www .planetark.com/avantgo/dailynewsstory.cfm?newsid=37623.

57 Greenpeace report on Guangdong drought; Elizabeth C. Economy, *The River Runs Black: The Environmental Challenge to China's Future* (Ithaca, NY: Cornell University Press, 2004), 6; "Drought, Floods Strike China, Affecting Tens of Millions," People's Daily Online, 12 May 2006, http://english.people .com.cn/200605/12/eng20060512_264969.html; satellite views at "Natural Hazards," NASA Earth Observatory, http:// earthobservatory.nasa.gov/NaturalHazards/natural_hazards _v2.php3?img_id=13559; "Water Flow in China's Yellow River Hits Record Low," Reuters, 7 Nov. 2006, available at www .planetark.com/dailynewsstory.cfm/newsid=38868/story.htm; Li Tianshe, "Headwaters of the Yellow River," at http://china .tyfo.com/int/travel/yellowriver/yellowriver.htm (source of quote); Mark Lynas, *High Tide: The Truth about Our Climate Crisis* (New York: Picador/St. Martin's Press, 2003), 138 and note; "Yellow River at Risk," Greenpeace/China, Beijing, 2005, plus additional information from Li Moxuan, Greenpeace expedition leader.

58 "Steel Giant Ordered to Halt Operation in Beijing by 2010," People's Daily Online, 26 Mar. 2005, http://english.people .com.cn/200503/26/eng20050326_178300.html; "Beijing's Main Polluter Ready to Move Out," *Asian Economic News,* 24 Dec. 2001, available at www.findarticles.com/p/articles/ mi_m0WDP/is_2001_Dec_24/ai_83370730; World Bank figures on twenty most polluted cities and other pollution levels quoted in Zijun Li, "Filthy Air Choking China's Growth, Olympic Goals," Worldwatch Institute, 14 Feb. 2006, www.worldwatch.org/node/3881; Honghong Yi et al., "Atmospheric Environmental Protection in China: Current Status, Developmental Trend, and Research Emphasis," *Energy Policy* 35 (Feb. 2007); Matt Walker, "A Nation Struggling to Catch Its Breath," *New Scientist* 190 (29 Apr. 2006),

www.newscientist.com/channel/earth/energy-fuels/
dn9082-china-struggling-to-catch-its-breath.html.

59 Economy, *River Runs Black,* 53–54, 64–66. Forest cover
estimate from www.unep-wcmc.org/forest/ofc_pan.htm.

60 "China Power," *OnEarth*/NRDC, winter 2004, www.nrdc
.org/OnEarth/04win/field.asp.

61 Peter Aldhous, "Energy: China's Burning Ambition,"
Nature 435 (30 June 2005); Shai Oster, "Illegal Power Plants,
Coal Mines in China Challenge for Beijing," *Wall Street Jour-
nal,* 27 Dec. 2006.

62 Economy, *River Runs Black,* 74, on cars; Howard French,
"A City's Traffic Plans Are Snarled by China's Car Culture,"
New York Times, 12 July 2005; Duncan Hewitt, "China's Environmental Challenge," BBC News Online, 17 Nov. 2000,
http://news.bbc.co.uk/1/hi/in_depth/asia_pacific/1027824
.stm; interview with Ding Yihui, a professor at the National
Climate Center, Beijing, at www.pbs.org/wgbh/nova/
worldbalance/voic-yihu.html; Keith Bradsher, "From China,
Some Relief on Oil Demand," *New York Times,* 14 July 2005;
"State of the World 2006," Worldwatch Institute, www.world
watch.org/node/3893.

63 "Annual Report 2006 (China)," Natural Resources Defense Council, at www.nrdc.org/about/annual/china.pdf;
Jessica Marshall, "Three Gorges Dam Threatens Vast Fishery," *New Scientist,* 25 Feb. 2006; Economy, *River Runs Black,*
184–85; and media sources, such as "China Shelves Twenty-
three Power Stations, Citing Environment," Reuters, 20 Jan.
2005, available at www.planetark.com/dailynewsstory.cfm/
newsid/29119/; Keith Bradsher, "China's Boom Adds to Global
Warming Problem," *New York Times,* 22 Oct. 2004; Aldhous,
"Energy: China's Burning Ambition"; Li, "Filthy Air Choking
China's Growth." Also see Yi, "Atmospheric Environmental
Protection in China."

64 Jim Yardley, "China Says Rich Countries Should Take
Lead on Global Warming," *New York Times,* 7 Feb. 2007.
Christopher L. Weber et al., "The Contribution of Chinese
Exports to Climate Change," *Energy Policy* 36 (Sept. 2008).

CHAPTER 5

1 *Bulletin of the Atomic Scientists,* Jan.–Feb. 2007 (the famous ticking clock, a metaphor for the threat of world anni-
hilation, created in 1947, was first set at 7 minutes to
midnight); James Hansen, "Defusing the Global Warming
Time Bomb," *Scientific American,* Mar. 2004. U.S. Senator
James Inhofe (R-OK) on the Senate floor on 28 July 2003, in
a speech titled "Catastrophic Global Warming Alarmism Not
Based on Objective Science" (full sentence is "I have offered
compelling evidence that catastrophic global warming is a
hoax"), available at http://inhofe.senate.gov/pressapp/record

.cfm?id=206907. Quotation is from James Gustave Speth, *Red
Sky at Morning,* rev. ed. (New Haven, CT: Yale University
Press, 2005), 204. The modern world had warning about the
climate effects of CO_2, beginning in 1957; see Spencer Weart,
The Discovery of Global Warming (Cambridge, MA: Harvard
University Press, 2003).

2 Andrew C. Revkin, "After Applause Dies Down, Global
Warming Talks Leave Few Concrete Goals," *New York Times,*
10 July 2008; Stephen Byers and Olympia Snowe, "Meeting
the Climate Challenge: Recommendations of the International Climate Change Task Force," Institute for Public Policy
Research, Center for American Progress, and Australia Institute, 2005.

3 I refer throughout this chapter to facts and figures stated
by the Intergovernmental Panel on Climate Change (IPCC)
in "Climate Change 2007: The Physical Science Basis. Summary for Policymakers," 2 Feb. 2007, and other Fourth Assessment Reports (AR4) available at www.ipcc.ch; published
research that formed the basis for the AR4; and scientist interviews. The figure of 1°F of ocean warming is from Bill
Curry of Woods Hole Oceanographic Institution, speaking at
the 11th Annual Symposium of the Stegner Center for Land,
Resources, and the Environment, University of Utah, Salt Lake
City, 4 Mar. 2006. CO_2 levels are from U. Siegenthaler et al.,
"Stable Carbon Cycle-Climate Relationship during the Late
Pleistocene," *Science* 310 (25 Nov. 2005); Thomas Stocker is
quoted on the rate of change in Richard Black, "CO_2 Highest
for 560,000 Years," *BBC News,* 24 Nov. 2005, at http://news
.bbc.co.uk/1/hi/sci/tech/4467420.stm. The phrase "almost
no possibility" is based on the IPCC framework of percentage
possibilities, whereby the chance of the CO_2 rise and associated atmospheric effects occurring without greenhouse gas
forcing is "extremely unlikely" (with a probability of less
than 5 percent); the change is "very likely" (a probability
greater than 90 percent) not due to natural causes alone.

4 Information on the IPCC scenarios is at www.grida.no/
climate/ipcc/emission.

5 Temperature increase from CO_2 additions is not instan-
taneous; there is a time lag owing to atmospheric processes
and the uptake of heat by the ocean. The figures I present
of possible future temperatures are scientists' forecasts based
on how the Earth's temperature has behaved until now and
what is known about ongoing greenhouse gas inputs into the
atmosphere. The possibility that warming will be greater due
to soil or ocean feedback is from Margaret Torn and John
Harte, "Missing Feedbacks, Asymmetric Uncertainties, and
the Underestimation of Future Warming," *Geophysical Research Letters* 33 (26 May 2006). Comments on the conservatism of some projections, including those of the IPCC,

include the warnings about much higher sea level rise expressed in chapter 1. See also A. Barrie Pittock, "Are Scientists Underestimating Climate Change?" *Eos* 87 (22 Aug. 2006). For technical discussions of projected temperature change, see IPCC, "Climate Change 2007."

6 James Hansen et al., "Global Temperature Changes," *Proceedings of the National Academy of Sciences* 39 (26 Sept. 2006). The 1992 climate treaty requires nations and scientists to assess the danger of increased climate change. The United States and 188 other nations that ratified the UNFCCC pledged to prevent "dangerous interference with climate." This is what the Kyoto Protocol is all about: it is a way to start guarding against dangerous climate change. See proceedings of the conference titled "Avoiding Dangerous Climate Change," Hadley Centre, Met Office, Exeter, Eng., 1–3 Feb. 2005, available at www.defra.gov.uk/environment/climatechange/internat/dangerous-cc.htm; also see Bryan K. Mignone et al., "Atmospheric Stabilization and the Timing of Carbon Mitigation," *Climatic Change* 88, nos. 3–4 (June 2008); and Hansen, "Defusing the Global Warming Time Bomb." John P. Holdren of Harvard University and the Woods Hole Research Center warns of present dangers at www.climatesciencewatch.org/index.php/csw/details/holdren_global_climate_disruption/.

7 For these and other severe climate events, see references above; "What Is Dangerous Climate Change?" Report of Symposium for UNFCCC, Meeting of COP-10, Buenos Aires, 14 Dec. 2004, available at www.european-climate-forum .net/pdf/ECF_beijing_results.pdf. Regarding methane deposits, see chapter 2 and Alexandra Witze, "Buried Deposits of Greenhouse Gases May Be More Unstable Than Thought," news@nature.com, 12 Dec. 2006, at www.nature.com/news/2006/061211/full/061211–6.html. UN humanitarian goals can be found at www.un.org/millenniumgoals; Jonathan Patz et al., "Impact of Regional Climate Change on Human Health," *Nature* 438 (17 Nov. 2005); and National Research Council, *Abrupt Climate Change: Inevitable Surprises,* U.S. National Academy of Sciences, National Research Council Committee on Abrupt Climate Change (Washington, DC: National Academies Press, 2002). Refugee figures are from Norman Myers, "Environmental Refugees: An Emergent Security Issue," 13th Economic Forum, Prague, 22 May 2005, available at www.osce.org/documents/eea/2005/05/14488_en .pdf. "Climate disruption" is from Holdren; NASA climate scientist Drew Shindell used the term "climate meltdown" in "Political Heat," *New York Times* magazine, 18 Feb. 2007.

8 Byers and Snowe, "Meeting the Climate Challenge"; in comparison with this and Hansen's 1.8°F redline (below), Mario Molina told a U.S. Senate committee on 21 July 2005 that added heating should be kept below 2.6°F—well under

the 560 ppm doubling of CO_2—at http://energy.senate.gov/public/index.cfm?FuseAction=Hearings.Testimony&Hearing_ID=1484&Witness_ID=4226. However, other scientists and technical observers warn that total CO_2 must remain below 450 ppm to avoid catastrophic change; this level is only thirty-five years away at the present rate of increase of 2 ppm a year; see "The Stern Review," *Ecologist*, 12 Jan. 2006. Quotation is from Millennium Ecosystem Assessment (preliminary draft), p. 15, available at www.millenniumassessment.org/en/index.aspx. See especially *Ecosystems and Human Well-Being: Synthesis,* Millennium Ecosystem Assessment (Washington, DC: Island Press, 2005), also available at www.millenniumassessment.org/en/Products.Synthesis.aspx; and James Lovelock, *The Revenge of Gaia: Earth's Climate Crisis and the Fate of Humanity* (New York: Basic Books, 2006).

9 Hansen, "Defusing the Global Warming Time Bomb"; James Hansen, "A Slippery Slope: How Much Global Warming Constitutes 'Dangerous Anthropogenic Interference'?" *Climatic Change* 68 (Mar. 2005); James Hansen et al., "Earth's Energy Imbalance: Confirmation and Implications," *Science* 308 (3 June 2005). Hansen et al., "Target Atmospheric CO_2: Where Should Humanity Aim?" available at www.columbia.edu/~jeh1/2008/TargetCO2_20080407.pdf.

10 Broecker, "The Biggest Chill," *Natural History*, Oct. 1987; "beast" quote is given by historian Spencer Weart at www.aip.org/history/climate/public2.htm. Mike Hulme, director of the esteemed Tyndall Centre for Climate Change Research in England, believes that "climate change is real, must be faced and action taken" but criticizes "the discourse of catastrophe." Hulme thinks words like "chaotic," "worse than we thought," and even "rapid" are not the language of science, arguing that "the IPCC scenarios of future climate change are significant enough"; *BBC News,* 4 Nov. 2006, at http://news.bbc.co.uk/1/low/sci/tech/6115644.stm.

11 Corinne Le Quéré, "How Much of the Recent CO_2 Increase Is Due to Human Activities?" 7 June 2005, www.realclimate.org/index.php?p=160. Each molecule of CO_2 contains two atoms of oxygen and an atom of carbon. Carbon's part of the total weight is about 27 percent; thus 30 billion tons of CO_2 contains about 8 billion tons of carbon. A shorthand way of referring to greenhouse emissions is simply as "carbon." However, because the important measure is how much a gas contributes to global warming, greenhouse emissions of all kinds are frequently given in CO_2 equivalents. This is the shorthand I will use. See Robert Service, "The Carbon Conundrum," *Science* 305 (13 Aug. 2005). Land use changes, which, like clouds, are imperfectly calculated by computer climate programs, are examined in Jonathan Foley et al., "Global Consequences of Land Use,"

Science 309 (22 July 2005); and Johannes Feddema et al., "The Importance of Land-Cover Change in Simulating Future Climates," *Science* 310 (9 Dec. 2005), with Commentary by Roger Pielke Sr., same issue. I used the following sources throughout this chapter for general figures on emissions and energy production (note that the numbers I report must be considered estimates; there are differing ways of reporting and computing emissions, and tallies are usually two to three years behind): IPCC, "Climate Change 2007: The Physical Science Basis," www.ipcc.ch; Energy Information Agency, at www.eia.doe.gov; Kevin A. Baumert, Timothy Herzog, and Jonathan Pershing, *Navigating the Numbers: Greenhouse Gas Data and International Climate Policy* (Washington, DC: World Resources Institute, 2005); "Key World Energy Statistics," International Energy Agency, 2006, at www.iea.org/w/bookshop/add.aspx?id=144; www.nationmaster.com; and http://earthtrends.wri.org. The most up-to-date tally of total CO_2 emissions is at the Global Carbon Project, www.globalcarbonproject.org/budget.htm. I thank Hermann Gucinski and Dan Lashof for computations on volume of carbon.

12 Car emission statistics from Robert Socolow, "Can We Bury Global Warming?" *Scientific American,* July 2005. Burning a gallon of gasoline makes 19.6 pounds of CO_2. U.S. energy use figures are from Energy Information Administration at www.eia.doe.gov/emeu/aer/pecss_diagram.html.

13 Barbara Freese, *Coal: A Human History* (Cambridge, MA: Perseus Books, 2003); Larry Gibson, interview, and aerial survey of mountaintop removal mining near Charleston, WV, by SouthWings, 18–19 Sept. 2005. Component pollutants are listed in Union of Concerned Scientists, "Environmental Impacts of Coal Power: Air Pollution," 18 Aug. 2005, at www.ucsusa.org/clean_energy/coalvswind/c02c.html; for mercury, see Janet Larsen, "Coal Takes Heavy Toll," Earth Policy Institute, 24 Aug. 2004, at www.earth-policy.org/Updates/Update42.htm. See also "The True Cost of Coal," Tri-State Citizens Mining Network, Washington, PA, 2004; John Vidal, "The End of Oil Is Closer Than You Think," *Guardian,* 21 Apr. 2005; "Oil-Demand Growth Conundrum," Trendvue, at http://trendvue.com/doc/11415. See also James Hansen and Makiko Sato, "Greenhouse Gas Growth Rates," *Proceedings of the National Academy of Sciences* 101 (16 Nov. 2004); Hansen, "Slippery Slope"; Jeremy Smith, "Oil and Security: Special Report," *Ecologist* 33 (Apr. 2003); Paul R. Epstein and Jesse Selber, eds., "Oil: A Life Cycle Analysis of Its Health and Environmental Impacts," Center for Health and the Global Environment, Harvard Medical School, 2002, at http://chge.med.harvard.edu/publications/documents/oilfullreport.pdf; Robert Hirsch, Roger Bezdek, and Robert Wendling, "Peaking of World Oil Production: Impacts, Miti-

gation, and Risk Management," Feb. 2005, available at www.netl.doe.gov/publications/others/pdf/Oil_Peaking_NETL.pdf. End-of-oil scenarios are played out in James Howard Kunstler, *The Long Emergency* (New York: Atlantic Monthly Press, 2005); Paul Roberts, *The End of Oil: On the Edge of a Perilous New World* (New York: Houghton Mifflin, 2004); www.hubbertpeak.com; and a website by Matt Savinar, www.lifeaftertheoilcrash.net. For oil's effect on developing nations, see, for example, Andrew Simms and Hannah Reid, "Africa—Up in Smoke 2: The Second Report from the Working Group on Climate Change and Development," New Economics International Foundation and the Institute for Environment and Development, 2005, pp. 20–21, available at www.iied.org/pubs/display.php?o=9560IIED.

14 Data from EPA, www.epa.gov/air/airtrends/index.html, and Toxic Release Inventory, interpreted by Physicians for Social Responsibility; Mark Clayton, "Why Coal-Rich U.S. Is Seeing Record Imports," *Christian Science Monitor,* 10 July 2006, www.csmonitor.com/2006/0710/p02s01-usec.html; "Department of Energy Tracks Resurgence of Coal-Fired Power Plants," DOE press release, 3 Aug. 2006, available at www.netl.doe.gov/publications/press/2006/06046-Coal-Fired_Power_Plants_Database.html; Timothy Gardner, "Planned U.S. Coal Plants Would Hike Warming—Group," Reuters, 24 July 2006, available at www.planetark.com/dailynewsstory.cfm/newsid/37367/newsDate/24-Jul-2006/story.htm. The EPA estimates 24,000 premature deaths from power plants; see J. R. Pegg, "Coal Power Soot Kills 24,000 Americans Annually," Environment News Service, 10 June 2004, www.ens-newswire.com/ens/jun2004/2004–06–10–10.asp. Coal could run out sooner; see Richard Heinberg, "Burning the Furniture," Mar. 2007, Global Public Media, http://globalpublicmedia.com/richard_heinbergs_museletter_179_burning the furniture.

15 Information on IGCC is from several energy sites, such as www.alliantenergy.com/docs/groups/public/documents/pub/p014448.hcsp. CO_2 sequestration is the subject of Klaus Lackner, "A Guide to CO_2 Sequestration," *Science* 300 (13 June 2003); see also IPCC, "Special Report on Carbon Dioxide Capture and Storage," 2005, at www.ipcc.ch/activity/srccs/index.htm; and David Hawkins et al., "What to Do About Coal," in "Energy's Future Beyond Carbon," *Scientific American* special issue (Sept. 2006); Darren Samuelsohn, "Underground Movement Emerges for Sequestering CO_2," *Greenwire,* 14 Aug. 2006, at www.eenews.net/Greenwire/2006/08/14/archive/1/?terms=sequestering.

16 U.S. Energy Information Administration, "Electric Power Annual," 4 Oct. 2006, at www.eia.doe.gov/cneaf/electricity/epa/epa_sum.html; Mark Clayton, "New Coal Plants Bury 'Kyoto,'" *Christian Science Monitor,* 24 Dec. 2004, www

.csmonitor.com/2004/1223/po1so4-sten.html; S. Pacala and R. Socolow, "Stabilization Wedges: Solving the Climate Problem for the Next Fifty Years with Current Technologies," *Science* 305 (13 Aug. 2004). Two new IGCC plants were planned as of January 2007, by GE and Hunton Energy; Andrew Ross Sorkin, "A Buyout Deal That Has Many Shades of Green," *New York Times,* 26. Feb. 2007. Coal can also be used to make synthetic gasoline, but the process known now, invented in Germany in the 1920s, is problematic because of high energy use and greenhouse gas output.

17 Henning Steinfeld et al., *Livestock's Long Shadow: Environmental Issues and Options* (Rome: Food and Agriculture Organization of the UN, 2006), available at www.virtualcentre.org/en/library/key_pub/longshad/A0701E00.htm.

18 Schneider quote is from http://stephenschneider.stanford.edu/Climate/Climate_Science/CliSciFrameset.html.

19 Stephen Schneider et al., "Costing Non-Linearities, Surprises, and Irreversible Events," *Pacific and Asian Journal of Energy* 10, no. 1 (2000); discussions at http://stephenschneider.stanford.edu/climate/climate_impacts/CliImpFrameset.html.

20 Broadly paraphrased from Jared Diamond, *Collapse* (New York: Viking-Penguin, 2005), 419–37; Thomas E. Lovejoy and Lee Hannah, "While Scientists Quibble, Species Vanish," *International Herald Tribune,* 15 Oct. 2005.

21 Notes from United Nations Climate Change Conferences COP11 and COP/MOP1, Montreal, 7–9 Dec. 2005, and COP/MOP12, Nairobi, 6–17 Nov. 2006; complete information on climate meetings is at http://unfccc.int/2860.php. Also U.S. Department of State, "President Bush and the Asia-Pacific Partnership on Clean Development," www.state.gov/g/oes/rls/fs/50314.htm and www.state.gov/r/pa/scp/2006/75384.htm; Ben Crystall, "The Big Clean-Up," *New Scientist,* 3 Sept. 2005.

22 "Along the Road from Kyoto," *Science* 311 (24 Mar. 2006), updated to 2002 or 2003 figures when possible from European Environment Agency and U.S. Environmental Protection Agency figures. The higher Canadian shortfall was reported at Jeff Sallot, "Kyoto Plan No Good, Minister Argues," *Globe and Mail,* 8 Apr. 2006, at www.theglobeandmail.com/servlet/story/LAC.20060408.PARLKYOTO08/TPStory/National; the Canadian government climate information site at www.climatechange.gc.ca/ was closed by the Harper government in 2006. See also "Climate Change Treaty One Year Old but Emissions Still on the Rise," Common Dreams Newswire, 15 Feb. 2006, www.commondreams.org/news2006/0215–06.htm.

23 Total radiative forcing in the atmosphere—the overall greenhouse effect—is calculated yearly by NOAA and reported at www.cmdl.noaa.gov/aggi/. See also Guus J. M. Velders et al., "The Importance of the Montreal Protocol in Protecting Climate," *Proceedings of the National Academy of Sciences* early edition (8 Mar. 2007); "Summing Up: Energy Symposium—'The Rosenfeld Effect,'" PowerPoint presentation, 28 Apr. 2006, available at www.energy.ca.gov/2006publications/CEC-999-2006-005/CEC-999-2006-005.PDF. The U.S. Department of Energy reported a 0.6 percent increase in fossil fuel CO_2 for 2005, a slowdown from 2004 that it attributed to higher energy prices; see www.eia.doe.gov/neic/press/press275.html; see Amory Lovins, "More Profit with Less Carbon," *Scientific American,* Sept. 2005.

24 Global Carbon Project, www.globalcarbonproject.org/carbontrendsindex.htm; IPCC, "Climate Change 2007"; Corinne LeQuéré, email, 31 Jan. 2007; "Trends in Atmospheric Carbon Dioxide," Earth System Research Laboratory, www.cmdl.noaa.gov/ccgg/trends; www.cmdl.noaa.gov/aggi.

25 Malte Meinshausen et al., "Multi-Gas Emission Pathways to Meet Climate Targets," *Climatic Change* 5 (Mar. 2006), and "How Much CO_2 Emission Is Too Much," 6 Nov. 2006, at www.realclimate.org/index.php/archives/2006/11/how-much-co2-emission-is-too-much/. The Global Commons Institute is at www.gci.org.uk.

26 An early and prescient paper comparing renewable energy with fossil fuel energy is David Pimentel et al., "Renewable Energy: Economic and Environmental Issues," *BioScience* 44 (Sept. 1994), available at http://dieoff.org/page84.htm; updated in "Renewable Energy: Current and Potential Issues," *BioScience* 52 (Dec. 2002), available at http://arec.oregonstate.edu/jaeger/energy/Renewable%20energy%20article%20pimental[*sic*].pdf. Throughout these pages I refer to information from Amory Lovins and the Rocky Mountain Institute (see notes below); Martin Hoffert et al., "Advanced Technology Paths to Global Climate Stability: Energy for a Greenhouse Planet," *Science* 298 (1 Nov. 2002); and "Energy's Future Beyond Carbon," *Scientific American* special issue (Sept. 2006). Also see "Sustainability and Energy," special section, *Science* 315 (9 Feb. 2007).

27 Nuclear power is one of the most controversial subjects in the energy field, and it is difficult to find unbiased general sources. The main sources I relied on are MIT, "The Future of Nuclear Power," 2003, http://web.mit.edu/nuclearpower/; Nuclear Energy Institute, "Nuclear Facts," www.nei.org/doc.asp?catnum=2&catid=106; Amory Lovins, "Nuclear Power: Economics and Climate-Protection Potential," Publication E05–08, 11 Sept. 2005, available at www.rmi.org/sitepages/pid171.php#LibNucEnergy; Eliot Marshall et al., "Is the Friendly Atom Poised for a Comeback?" *Science* 309 (19 Aug.

2005); and Nuclear Information and Resource Service, www
.nirs.org/home.htm. John Christy quote is from an email of
9 Mar. 2006.

28 Information on French reactors during the 2003 heat
wave is from UN Environment Programme, "Impacts of
Summer 2003 Heat Wave in Europe," GRID-Europe, Mar.
2004, www.grid.unep.ch/product/publication/download/
ew_heat_wave.en.pdf.

29 "Incidents" are documented in Greenpeace, "An American Chernobyl: Nuclear 'Near Misses' at U.S. Reactors since
1986," 21 Apr. 2006, available at www.greenpeace.org/usa/
press/reports/an-american-chernobyl-nuclear. See also
William H. Hannum et al., "Smarter Use of Nuclear Waste,"
Scientific American, Dec. 2005; and various articles on the
Chemical and Engineering News site, such as Michael Free-
mantle, "Nuclear Power for the Future," http://pubs.acs.org/
cen/coverstory/8237/8237nuclearenergy.html.

30 New reactor designs are explained somewhat technically
at www.eia.doe.gov/cneaf/nuclear/page/analysis/nucenviss2
.html.

31 The amount of usable solar energy, moderated of course
by clouds and dust, depends on sun angle and length of day
and averages about 342 watts per square meter. Possible wind
generator sites worldwide could generate thirty-six times the
electrical energy used by the world, which is about 2 terawatts
(2×10^{12}), according to Christina Archer and Mark Jacobson,
"Evaluation of Global Wind Power," *Journal of Geophysical
Research* 110 (30 June 2005). Total world energy consumption
from oil, coal, etc. is about 14 terawatts. General energy figures are from "Renewables Global Status Report, 2006 Update," Renewable Energy Policy Network, Paris, available at
www.ren21.net; Eric Martinot, "Global Revolution: A Status
Report on Renewable Energy Worldwide," *Earthscan,* 2 Nov.
2005, www.earthscan.co.uk/news/article/mps/uan/508/v/3/
sp/; U.S. research into renewable energy is discussed at www
.nrel.gov/about.html; and a very readable review of U.S.
nonfossil energy sources is Worldwatch Institute and Center
for American Progress, "American Energy: The Renewable
Path to Energy Security" (Sept. 2006), available at www
.americanenergynow.org.

32 On German energy, see Michael Levitin, "Nein Lives,"
Grist, 12 Aug. 2005, www.grist.org/news/maindish/2005/
08/12/levitin-germany/; "German Wind Energy Report Opens
Heated Debate," EurActiv, 2 Mar. 2005, www.euractiv.com/
en/energy/german-wind-energy-report-opens-heated-debate/
article-136239; Jürgen Trittin, "Germany's New Climate
Change Program," address at the Hague, 22 Nov. 2000, www
.bmu.de/english/climate_change/doc/pdf/3314.pdf.

33 Lester Brown, "Wind Energy Demand Booming," Earth

Policy Institute, 22 Mar. 2006, www.earth-policy.org/Updates/
2006/Update52.htm. Wind power information may be accessed at www.awea.org and www.ewea.org (American and
European Wind Energy Association websites, respectively);
and information warning of its dangers may be found at
www.aweo.org, www.windwatch.org, and www.batcon.org;
I also spoke to Appalachian wind farm opponent Dan Boone
by telephone. See Arthur O'Donnell, "Altamont Wind Ops
Trimmed to Cut Raptor Kills," Landletter, 22 Feb. 2007, at
www.eenews.net/landletter/2007/02/22. USFWS interim wind
siting guidelines are at www.fws.gov/habitatconservation/
wind.pdf. Along with the older "erector set" type of wind
machines, there are more than 85,000 tall communications
towers that may be a much larger problem for migrating
birds; see www.abcbirds.org/policy/towerkill.htm.

34 Oregon Health Sciences University building information
at www.ohsu.edu/ohsuedu/newspub/releases/021606greenest
.cfm; "Manchester Gets Largest Solar Tower in Europe," 13
May 2005, at www.edie.net/news/news_story.asp?id=9906.

35 Hoffert et al., "Advanced Technology Paths" (2002), updated in Michael Parfit, "Future Power," *National Geographic,*
Aug. 2005. By one estimate, 61,000 square miles (158,000 sq
km) are paved over with roads and parking lots; see Lester
Brown, "Paving the Planet: Cars and Crops Competing for
Land," 14 Feb. 2001, www.earth-policy.org/Alerts/Alert12
.htm. See also "Energy Foundation Study Finds Residential
and Commercial Rooftops Could Support Vast U.S. Market
for Solar Power," Power On Line, 1 Mar. 2005, www.power
online.com/content/news/article.asp?docid=7137b00b-41db-
4fba-af73-29ec4c958eea. Installation information from Janet
McGarry, "Nation's Largest Rooftop Solar Installation,"
www.nesea.org/publications/NESun/largest_installation
.html; Riva Richmond, "Google Plans to Build Huge Solar
Energy System for Headquarters," *MarketWatch,* 17 Oct.
2006, www.marketwatch.com/News/Story/Story.aspx?dist=
newsfinder&siteid=mktw&guid=%7B630082C7-5370-46C2
-8041-FC3FCA28CA16%7D.

36 Ovshinsky information is from the website for "Scientific American Frontiers," www.kpbs.org/saf/1506/features/
ovshinsky.htm.

37 IEEE-USA, "Solar and Other Renewable Energy Technol-
ogies," www.ieeeusa.org/policy/positions/solar.html; infor-
mation on the Kramer Junction Solar Electric Generating
Station can be found in land use data base of the Center for
Land Use Interpretation, http://ludb.clui.org/ex/i/CA9679;
"Sandia, Stirling to Build Solar Dish Engine Power Plant," 9
Nov. 2004, www.sandia.gov/news/resources/releases/2004/
renew-energy-batt/Stirling.html; Emma Graham-Harrison,
"Energy-Hungry China Warms to Solar Water Heaters,"

Reuters, 5 June 2006, available at www.planetark.org/daily
newsstory.cfm?newsid=36636.

38 Nathan Lewis and Daniel Nocera, "Powering the Planet:
Chemical Challenges in Solar Energy Utilization," *Proceedings
of the National Academy of Sciences* 103 (24 Oct. 2006).

39 On ocean and tidal power, see Jeff Johnson, "Power from
Moving Water," *Chemical and Engineering News,* 4 Oct. 2004,
http://pubs.acs.org/cen/coverstory/8240/8240energy.html;
on New York tidal power, see www.verdantpower.com/
initiatives/currentinit.html; Carolyn Elefant and Sean O'Neill,
"Ocean Energy Report for 2005," 9 Jan. 2006, http://renew
ableenergyaccess.com/rea/news/story?id=41396.

40 Ping Xie, "Three-Gorges Dam: Risk to Ancient Fish,"
Science 302 (14 Nov. 2003); Guozhen Shen and Zongqiang
Xie, "Three Gorges Project: Chance and Challenge," *Science*
304 (30 Apr. 2004); and Jessica Marshall, "Three Gorges Dam
Threatens Vast Fishery," *New Scientist,* 25 Feb. 2006. Tropical
dams may be a large source of CO_2 and methane, according
to Patrick McCully, "Flooding the Land, Warming the Earth,"
International Rivers Network (2002) at www.irn.org/
programs/greenhouse/frontpage.

41 "Geothermal Power Issue Brief," Renewable Energy Policy Project, Dec. 2003, available at www.repp.org/geothermal/
index.

42 Daniel M. Kammen, "Cookstoves for the Developing
World," at http://socrates.berkeley.edu/~kammen/cook
stoves.html (originally published in *Scientific American,* July
1995; contains references to World Bank paper).

43 NRDC, "Growing Energy," available at www.nrdc.org/
air/energy/biofuels/contents.asp. "Biofuels for Transportation," Worldwatch Institute, Washington, DC, 7 June 2006,
also contains a review of Brazil's ethanol industry.

44 Renewable Fuels Association, www.ethanolrfa.org/
industry/#B; NRDC, "Ethanol: Energy Well Spent," Feb.
2006, available at www.nrdc.org/air/transportation/ethanol
/ethanol.asp; Patrick DiJusto, "Blue-Green Acres," *Scientific
American,* 29 (Sept. 2005).

45 For bioplastics, see Joel Makower, "Kernels of Hope for
Bioplastics," www.worldchanging.com/archives/003535.html;
Arthur Ragauskas et al., "The Path Forward for Biofuels and
Biomaterials," *Science* 311 (27 Jan. 2006). For biodiesel, see
www.biodieselamerica.org/what_is_biodiesel; for an agricul-
tural view on energy generally, see www.agmrc.org/agmrc/
commodity/energy; Patrick Barta and Jane Spencer, "Crude
Awakening: As Alternative Energy Heats Up, Environmental
Concerns Grow," *Wall Street Journal,* 5 Dec. 2006; also see
Lorien Holland, "Whose Forest?" *Newsweek* (international
edition), 2 Jan. 2006, available at www.msnbc.msn.com/
id/10510089/site/newsweek.

46 Cutler J. Cleveland, "Net Energy from the Extraction of
Oil and Gas in the United States," *Energy* 30 (5 Apr. 2005);
Alexander Farrell et al., "Ethanol Can Contribute to Energy
and Environmental Goals," *Science* 311 (27 Jan. 2006); Nathan
Glasgow and Lena Hansen, "Setting the Record Straight on
Ethanol," Rocky Mountain Institute, fall 2005, www.rmi
.org/sitepages/pid1157.php; Center for American Progress,
"Resources for Global Growth: Agriculture, Energy, and
Trade in the 21st Century," 2005, www.americanprogress
.org/site/pp.asp?c=biJRJ8OVF&b=1241501; Jason Hill et al.,
"Environmental, Economic, and Energetic Costs and Benefits
of Biodiesel and Ethanol Biofuels," *Proceedings of the National
Academy of Sciences* 103 (25 July 2006); David Pimentel and
Tad W. Patzek, "Ethanol Production Using Corn, Switch-
grass, and Wood; Biodiesel Production Using Soybean and
Sunflower," *Natural Resources Research* 14 (Mar. 2005); Roel
Hammerschlag, "Ethanol's Energy Return on Investment: A
Survey of the Literature 1990–Present," *Environmental Science
and Technology* 40 (15 Mar. 2006); Timothy Searchinger
et al., "Use of U.S. Croplands for Biofuels Increases Green-
house Gases through Emissions from Land-Use Change,"
Science 319 (29 Feb. 2008); Joseph Fargione et al., "Land
Clearing and the Biofuel Carbon Debt," *Science* 319 (29 Feb.
2008).

47 Lester Brown, "Distillery Demand for Grain to Fuel Cars
Vastly Understated," Earth Policy Institute, 4 Jan. 2007, at
www.earth-policy.org/Updates/2007/Update63.htm; "Biofuels for Transportation," Worldwatch Institute, Washington
DC, 7 June 2006, at www.bioenergy-world.com/americas/
2006/IMG/pdf/Biofuels_for_Transport_Worldwatch_
Institute.pdf. Information on "25 by 25" is available at www
.25x25.org. For a view emphasizing food first and regional
self-sufficiency, see Mae-Wan Ho et al., "Which Energy?,"
Institute of Science in Society (London), 2006, at www.i-sis
.org.uk/onlinestore/books.php#238. Also see Richard A. Kerr
and Robert F. Service, "What Can Replace Cheap Oil—and
When?" *Science* 309 (1 July 2006).

48 Joseph Romm, "The Hype about Hydrogen," *Issues in
Science and Technology* (Spring 2004), at www.issues.org/
20.3/romm.html; John Holdren, "The Energy Innovation
Imperative," *Innovations* (Spring 2006), available at www
.mitpressjournals.org/doi/pdf/10.1162/itgg.2006.1.2.3; Steven
Ashley, "On the Road to Fuel-Cell Cars," *Scientific American*
(March 2005); www.greenhydrogencoalition.org. Encouraging views may be found at "Twenty Hydrogen Myths," www
.rmi.org/sitepages/art7516.php, www.hydrogennow.org, and
"Hydrogen Economy Fact Sheet" at www.whitehouse.gov/
news/releases/2003/06/20030625–6.html. There are other
technologies and so-called "geo-engineering" fixes for energy

and global warming that are more or less impractical now but have advocates. See W. Wayt Gibbs, "Plan B for Energy," *Scientific American* special issue (Sept. 2006); Jesse Ausubel, "Big Green Energy Machines," *Industrial Physicist* (Oct.–Nov. 2004); William Broad, "How to Cool a Planet (Maybe)," *New York Times,* 27 June 2006, available at www.nytimes.com/2006/06/27/science/earth/27cool.html?ex=1309060800&en=dod351a5cf6b48d1&ei=5088&partner=rssnyt&emc=rss.

49 Information on the rise of the SUV is in Malcolm Gladwell, "Big and Bad," *New Yorker,* 12 Jan. 2004, available at www.gladwell.com/2004/2004_01_12_a_suv.html; and Keith Bradsher, *High and Mighty: SUVs—The World's Most Dangerous Vehicles and How They Got That Way* (New York: PublicAffairs, 2002). Vehicle CO_2 numbers are from "Greenhouse Gas Emissions from the U.S. Transportation Sector, 1990–2003," Environmental Protection Agency, 2006, www.epa.gov/OMS/climate/420r06003.pdf.

50 Kristian Bodek, MIT Energy Club, "U.S. Transportation Fact Sheet," 18 Oct. 2006, at http://web.mit.edu/mit_energy/resources/factsheets/TransportationUS.pdf; David Friedman, Jessica Biegelson, and Scott Nathanson, "Shopping for a Hybrid? Buyer Beware," *Catalyst* 4, no. 2 (Fall 2005), at www.ucsusa.org/publications/catalyst/shopping-for-a-hybrid.html; John O'Dell, "Vehicle Mileage Estimates Get Real," *Newsday,* 12 Dec. 2006, at www.newsday.com/news/nationworld/nation/la-fi-fuel12dec12,0,2833292.story?coll=ny-top-headlines; Lester Brown, "The Short Path to Oil Independence," Earth Policy Institute, 13 Oct. 2004, www.earth-policy.org/Updates/Update43.htm; Jim Motavalli, "Cleaner, Greener Cars," *E* magazine, Mar.–Apr. 2007. Seventeen million new cars and light trucks are bought each year in the United States, and according to figures in the Transportation Energy Data Book, at http://cta.ornl.gov/data/chapter3.shtml, the median life of all vehicles is seventeen years. FedEx and UPS are beginning to use trucks with hybrid-electric and hybrid-hydraulic technology, respectively.

51 Auto efficiency estimates from www.40mpg.org and NRDC; Persian Gulf Oil and Gas Exports Fact Sheet, Sept. 2004, available at www.eia.doe.gov/emeu/cabs/pgulf.html. See also "Gasoline and the American People," at www2.cera.com/gasoline/summary/.

52 Environmental Defense, "Stalled on Carbon Pollution," 9 Aug. 2005, www.environmentaldefense.org/article.cfm?contentID=4721; car efficiency data from Sheryl Carter, NRDC, speaking at the 11th Annual Symposium of the Stegner Center for Land, Resources, and the Environment, University of Utah, Salt Lake City, 4 Mar. 2006; composite-material car concept at www.rmi.org/sitepages/pid386.php. To their credit, Ford has an office building and GM an assembly plant

that are certified green buildings; see "GM Opens First-Ever LEED-Certified Auto Plant," at www.greenbiz.com/sites/greenerbuildings/news_detail.cfm?NewsID=33486.

53 Richard Register, *Ecocities* (Gabriola Island, B.C.: New Society, 2006); and "Green Cities," *Common Ground* (June 2005).

54 Fred Pearce, "Ecopolis Now," *New Scientist* (17 June 2006); and *Fragile Earth: Views of a Changing World* (London: Collins, 2006).

55 Paul Hawken et al., *Natural Capitalism* (Boston: Little, Brown, 1999); interview with Register, Berkeley, CA, 15 Sept. 2006; Pearce, "Ecopolis Now"; see www.ecocitybuilders.org.

56 Public transportation figures are available at www.publictransportation.org/reports/asp/air.asp; Michael Parfit, "Future Power," *National Geographic* (Aug. 2005). Rail savings figures are from Register, www.ecocitybuilders.org.

57 For an interview with Jaime Lerner, see Reed McManus, "Imagine a City with 30 Percent Fewer Cars," *Sierra* (Jan./Feb. 2006), www.sierraclub.org/sierra/200601/interview.asp.

58 For LEED and Building America, see www.usgbc.org and www.eere.energy.gov/buildings/building_america/about.html. Auden Schendler and Randy Udall advocate improvements to LEED: "LEED Is Broken; Let's Fix It," Grist, 26 Oct. 2005, www.grist.org/comments/soapbox/2005/10/26/leed/index1.html; Katie Zezima, "Boston Plans to Go 'Green' on Large Building Projects," *New York Times,* 20 Dec. 2006. Factsheets "Residential Buildings" and "Commercial Buildings," University of Michigan Center for Sustainable Systems, at http://css.snre.umich.edu/facts. On net-zero prefab houses see Eberhard Jochem, "An Efficient Solution," *Scientific American* special issue (Sept. 2006).

59 For a list of "climate protection cities," see http://www.iclei.org/index.php?id=800. The U.S. Mayors' Climate Protection Agreement is at www.ci.seattle.wa.us/mayor/climate and includes a list of twelve specific actions for cities to undertake; Seattle's list of actions is available at www.seattle.gov/climate/report.htm; City of Portland Office of Sustainable Development, "Global Warming Progress Report," June 2005, is available at www.portlandonline.com/osd/. See also the World Mayors website at www.citymayors.com. On green roofs, see www.greenroofs.net, and for news of local energy programs and laws, refer to www.pewclimate.org/what_s_being_done/in_the_states/news.cfm; and "Points of Light," *E* Magazine (July–Aug.) 2006.

60 Speth, *Red Sky,* 68–69; State Attorneys General, Letter to President Bush, July 17, 2002, reported in the *New York Times* in an article available at www.commondreams.org/headlines02/0717–01.htm. A comprehensive list of local initiatives is available from Civil Society Institute, "Politics Abhors a Vacuum: Regional, State, Municipal, and Corporate Action on Climate Change," at www.resultsforamerica.org/calendar/files/ClimateChangeActionOverview.pdf; and www.pewclimate.org/what_s_being_done/in_the_states. For the northeastern states' initiative, see www.rggi.org.

61 For information on the new California law, see the Union of Concerned Scientists' statement at www.ucsusa.org/news/press_release/california-enacts-nations.html; for California climate action, see www.climatechange.ca.gov/index.html.

62 On energy use and losses, see "U.S. Energy Flow Trends 2002," http://eed.llnl.gov/flow/02flow.php. For Lovins, see next note.

63 Information and quotes in this section are from Amory Lovins, "More Profit, Less Carbon," *Scientific American* (Sept. 2005); and Lovins et al., "Winning the Oil Endgame," Executive Summary, Rocky Mountain Institute, Sept. 2004, available at www.rmi.org/store/p12details4772.php (the book can be downloaded at www.oilendgame.com/ReadTheBook.html). A study by the American Solar Energy Society calculated that more than half of the reductions needed to bring U.S. energy use down by 80 percent by 2030 could come from increased efficiency; see "Tackling Climate Change in the U.S." at www.ases.org/climatechange.

64 BP's emissions reduction actions are listed in response to the Carbon Disclosure Project at www.cdproject.net/download.asp?file=CDP2_BP_response_2559.doc. BP remains the oil company with the most public pronouncements about the environment; see, for example, "BP and Climate Change," www.bp.com/subsection.do?categoryId=4529&contentId=7014604. For outside views, see Darcy Frey, "How Green Is BP?" *New York Times*, 8 Dec. 2002, available at www.mindfully.org/Industry/BP-How-Green8dec02.htm; and "BP," SourceWatch, www.sourcewatch.org/index.php?title=BP, for a recent chronology of newsworthy events. ExxonMobil has a new CEO and recently acknowledged the role of fossil fuels in global warming, but see an analysis of its corporate governance regarding climate at www.ceres.org/pub/publication.php?pid=96. On paid skeptics and obstructionists, see Union of Concerned Scientists, "Smoke, Mirrors, and Hot Air: How ExxonMobil Uses Big Tobacco's Tactics to 'Manufacture Uncertainty' on Climate Change," 2006, at www.ucsusa.org/news/press_release/ExxonMobil-GlobalWarming-tobacco.html (links to report on page); www.motherjones.com/news/feature/2005/05/some_like_it_hot.html; and Ross Gelbspan, *Boiling Point: How Politicians, Big Oil and Coal, Journalists, and Activists Are Fueling the Climate Crisis—And What We Can Do to Avert Disaster* (New York: Basic Books, 2004).

65 Carbon Disclosure Project's website is www.cdproject.net.

66 Ceres is at www.ceres.org, where the March 2006 report "Corporate Governance and Climate Change: Making the Connection" is available. "Managing the Risks and Opportunities of Climate Change: A Practical Toolkit for Corporate Leaders," Ceres, Boston, Jan. 2006, available at www.greenbiz.com/toolbox/tools_third.cfm?LinkAdvID=67111. Chevron's emissions inventory protocol may be seen at www.chevron.com/social_responsibility/environment/docs/2003_emissions_protocol.pdf; see also www.chevron.com/social_responsibility/environment/ghg_audit_response.asp for an "independent review of Chevron Texaco's greenhouse gas emissions inventory." For the Senate hearing report see Amanda Griscom Little, "Cap of Good Hope," Grist, 6 Apr. 2006, www.grist.org/news/muck/2006/04/06/griscom-little/index.html. Corporate/NGO coordinated energy plans include the Climate Action Partnership, at www.us-cap.org. For an environmental view of Chevron, BP, and other companies, see "Pick Your Poison: An Environmentalist's Guide to Gasoline," *Sierra* (Sept.–Oct. 2001), www.sierraclub.org/sierra/200109/hattam.asp.

67 Carbon trading information is from Karan Capoor and Philippe Ambrosi, "State and Trends of the Carbon Market 2006," World Bank/International Emissions Trading Association, Oct. 2006, at www.carbonfinance.org/docs/StateoftheCarbonMarket2006.pdf; and www.defra.gov.uk/environment/climatechange/trading/index.htm. Chicago Climate Exchange information is from www.ieta.org/ieta/www/pages/index.php?idSitePage=25. This exchange may have loopholes and not always insure that climate projects would not otherwise have come about; see NRDC's Dale Bryk at www.plentymag.com/features/2006/08/offset_upset.php. Also see David Greising, "The Carbon Frontier," *Bulletin of the Atomic Scientists* (Jul.–Aug. 2008).

68 The distribution of Clean Development Mechanism projects is shown at http://cdm.unfccc.int/Statistics/Registration/NumOfRegisteredProjByHostPartiesPieChart.html. See also "Chemicals Trapped between Treaties Present Serious Threat to Climate and Ozone Layer," Environmental Investigation Agency, press release (10 Nov. 2006); Keith Bradsher, "Big Profits, and Questions, in Effort to Cut Emissions," *New York Times*, 21 Dec. 2006.

69 For these companies, see www.kpcb.com/initiatives and www2.goldmansachs.com/our_firm/our_culture/corporate_citizenship/environmental_policy_framework/index.html; Matt Richtel, "Tech Barons Take On New Project: Energy Policy," *New York Times*, 29 Jan. 2007; "Tilting At Windmills,

Economist, 18 Nov. 2006. For more on companies with pro-active greenhouse gas reduction programs, see information provided by the Climate Group at www.theclimategroup .org/index.php?pid=373; World Wildlife Fund at www .worldwildlife.org/climate/projects/climateSavers.cfm; and by the Pew Center on Global Climate Change at www.pew climate.org/companies_leading_the_way_belc/company_ profiles. On corporate greenwashing and media spin, see the Center for Media and Democracy's SourceWatch, at www .sourcewatch.org/index.php?title=SourceWatch.

70 "Air Travel Heats Up the Planet," Sightline Institute, Seat-tle, Aug. 2004, www.sightline.org/research/energy/res_pubs/ rel_air_travel_aug04; H. L. Rogers et al., "The Impacts of Aviation on the Atmosphere," QinetiQ, Feb. 2003, www .ozone-sec.ch.cam.ac.uk/EORCU/Files/Rogers1.pdf; Mark Barrett, Pollution Control Strategies for Aircraft," World Wide Fund for Nature, Gland, Switz., 1994.

71 Car vs. airliner figures are 0.9 pound of CO_2 per mile for a car driver vs. 0.65 pound for the average U.S. air passenger, from "Transportation Greenhouse Gas Emissions," Rocky Mountain Institute, www.rmi.org/sitepages/pid342.php; higher figures are at "Air Travel Heats Up the Planet," Sight-line Institute, and in CE Delft comparisons quoted in "The Sky's the Limit," *Economist,* 10 June 2006; "Development of World Scheduled Revenue Traffic," International Civil Avia-tion Organization database of Total Services, provided by email by Attilio Costaguta, 19 Apr. 2006; Honor Mahony "U.S., EU Square Off on Airline Pollution," 5 Dec. 2006, at www.businessweek.com/globalbiz/content/dec2006/ gb20061205_157068.htm. For ships, see V. Eyring et al., "Multi-model Simulations of the Impact of International Shipping on Atmospheric Chemistry and Climate in 2000 and 2030," *Atmospheric Chemistry and Physics Discussions* 6 (12 Sept. 2006).

72 Pimentel, "Renewable Energy: Current and Potential Issues"; Holdren, "The Energy Innovation Imperative." For information on carbon offsets, which some think of as a use-ful substitute for conservation and direct emissions reduc-tions but which others describe as a modern day selling of indulgences, see the References.

73 Information on home and small business energy savings is available on most utility and environmental group websites (see References). See the American Council for an Energy-Efficient Economy, www.aceee.org/consumer/consumer .htm; Energy Star, a joint program of the U.S. EPA and Department of Energy, www.energystar.gov; and Rocky Mountain Institute at www.rmi.org/sitepages/pid16.php. Also see the McKinsey Global Institute report "Productiv-ity of Growing Global Energy Demand: A Microeconomic

Perspective," Nov. 2006, at www.mckinsey.com/mgi/publica tions/Global_Energy_Demand/index.asp. For power plant information, http://carma.org.

74 Information on agricultural energy use is from Sustain-able Table's "Fossil Fuel and Energy Use," www.sustain abletable.org/issues/energy, which references Leo Horrigan, Robert S. Lawrence, and Polly Walker, "How Sustainable Agriculture Can Address the Environment and Human Health Harms of Industrial Agriculture," *Environmental Health Perspectives* 110 (5 May 2002); Danielle Murray, "Oil and Food: A Rising Security Challenge," Earth Policy Institute, 9 May 1995, www.earth-policy.org/Updates/2005/ Update48_printable.htm; Rich Pirog and Andrew Benjamin, "Checking the Food Odometer," Leopold Center for Sus-tainable Agriculture, Iowa State University, 2003, www.leopold .iastate.edu/pubs/staff/files/food_travel072103.pdf; Gideon Eshel and Pamela A. Martin, "Diet, Energy, and Global Warming," *Earth Interactions* 10 (Apr. 2006); Melanie Warner, "Wal-Mart Is Going Organic, and Brand Names Get in Line," *New York Times,* 12 May 2006; David Pimentel et al., "Organic and Conventional Farming Systems: Environmental and Economic Issues," *Bioscience* 55 (July 2005); Henning Steinfeld et al., *Livestock's Long Shadow.* Rich references for organic and permaculture agriculture are Andrew Kimbrell, ed., *Fatal Harvest: The Tragedy of Industrial Agriculture* (Wash-ington, DC: Island Press/Foundation for Deep Ecology, 2002); and Bill Mollison, *Introduction to Permaculture* (Sisters Creek, Tasmania: Tagari, 1997).

75 *Stern Review on the Economics of Climate Change,* 30 Oct. 2006, is available at www.hm-treasury.gov.uk/independent _reviews/stern_review_economics_climate_change/sternreview _index.cfm. World GDP figures are available at https://www .cia.gov/library/publications/the-world-factbook/rankorder/ 2001rank.html.

76 Tol is quoted in "Stern Warning," *Economist,* 4 Nov. 2006. William Nordhaus, "Geography and Macroeconomics: New Data and New Findings," *Proceedings of the National Academy of Sciences* 103 (7 Mar. 2006), and Nordhaus, "The *Stern Review* on the Economics of Climate Change," 17 Nov. 2006, available at http://nordhaus.econ.yale.edu/SternReviewD2.pdf. For an-other review of actions and costs in controlling emissions, see John Hawksworth, "The World in 2050: Impact of Global Growth on Carbon Emissions and Climate Change Policy," a study produced for PriceWaterHouseCoopers, available at http://www.pwc.com/extweb/pwcpublications.nsf/docid/dfb54 c8aad6742db852571f5006dd532. Also see Christian Azar and Stephen Schneider, "Are the Economic Costs of Stabilising the Atmosphere Prohibitive?," *Ecological Economics* 42 (Aug. 2002). An interesting review of other economic estimates of global warming costs and critiques is Kate Galbraith, "The Trillion Dollar Question," Grist, 16 Nov, 2006, at www.grist.org/news/maindish/2006/11/16/galbraith.

77 Sheikh Yamani is quoted by Lovins in "Winning the Oil Endgame"; the survey is Hoffert et al., "Advanced Technology Paths."

78 Lovins, "More Profit, Less Carbon." A list of U.S. greenhouse gas research and reduction programs is at www.whitehouse.gov/news/releases/2005/05/20050518–4.html; national renewable programs can be found at www.nrel.gov; see also Renewable Energy Policy Project, www.repp.org/index.html. A review of subsidies is in Norman Myers and Jennifer Kent, *Perverse Subsidies: How Misused Tax Dollars Harm the Environment and the Economy* (Washington DC: Island Press, 2001); the transaction tax and other important solutions are in Gelbspan, *Boiling Point,* 182ff. A list of what executive and legislative branches could do in short order, presented in a partisan way but nevertheless a good template for any national leader, is "America Is Addicted to Oil: Ten Tough Questions and Answers for President Bush on Kicking the Oil Habit," report by the Center for American Progress, Feb. 2006, available at www.americanprogress.org/issues/2006/02/b1408771.html.

79 Pacala and Socolow, "Stabilization Wedges"; see also David Doniger et al., "An Ambitious, Centrist Approach to Global Warming Legislation," *Science* 314 (2 Nov. 2006); and Interlaboratory Working Group, "Scenarios for a Clean Energy Future," Nov. 2000, at www.ornl.gov/sci/eere/cef. For a list of sources of plans for climate actions, see References.

80 Daniel Kammen, "The Rise of Renewable Energy," *Scientific American* special issue (Sept. 2006); Andrew Revkin, "Budgets Falling in Race to Fight Global Warming," *New York Times,* 30 Oct. 2006; Beth Daley, "NASA Shelves Climate Satellites," *Boston Globe,* 9 June 2006, www.boston.com/news/nation/articles/2006/06/09/nasa_shelves_climate_satellites/; Jesse Ausubel, "Technical Progress and Climatic Change," *Energy Policy* 23, nos. 4–5 (1995). For a report on government interference with scientific information, see

Timothy Donaghy et al., "Atmosphere of Pressure," Union of Concerned Scientists, Feb. 2007, available at www.ucsusa.org/publications.

81 Kennedy speech at Rice University, Houston, TX, 12 Sept. 1962, available in print transcript and as audio at www1.jsc.nasa.gov/er/seh/ricetalk.htm.

82 Henry N. Pollock, *Uncertain Science . . . Uncertain World* (Cambridge: Cambridge University Press, 2003); skeptical views and serious questions on climate science are asked and responded to on www.realclimate.org; Spencer Weart, "Changing the Climate . . . of Public Opinion," *APS News,* Feb. 2006, www.aps.org/apsnews/0206/020614.cfm; also see his website on the history of global warming science, www.aip.org/history; IPCC, "Climate Change 2007"; interview with Parmesan, Earth & Sky radio series, 20 May 2002, www.earthsky.org/radioshows/45836/parmesan-i. Views of some scientists questioning the consensus have been mentioned in other sections of the book.

83 Molina and Toepher, comments at Stony Brook World Environmental Forum, State University of New York at Stony Brook, 6–8 May 2005.

84 Oil cost calculation is from Center for American Progress, "Resources for Growth." Twenty-nine percent of 2005 U.S. deficit was for petroleum products, per *New York Times,* 10 Feb. 2006; www.iht.com/articles/2006/02/10/business/usecon.php.

85 Donald A. Brown, "The Ethical Dimensions of Environmental Issues," *Daedalus* (Fall 2001).

86 Register, *Ecocities;* Toepher at Stony Brook conference, 8 May 2005.

87 Diamond, *Collapse,* 485; Tim Wirth, Steering Committee Chairman, Energy Future Coalition, quoted in J. R. Pegg, "Group Aims to Take Politics Out of Energy Policy," Environment News Service, 18 June 2003, www.ens-newswire.com/ens/jun2003/2003–06–18–10.asp; address by Edward O. Wilson at the Cornell Lab of Ornithology's "Science Takes Flight" Celebration, Ithaca, NY, 4 Sept. 2003.

REFERENCES

BOOKS ON CLIMATE CHANGE

Abbasi, Daniel R. *Americans and Climate Change: Closing the Gap between Science and Action.* New Haven: Yale University Press, 2006.

Alley, Richard B. *The Two-Mile Time Machine: Ice Cores, Abrupt Climate Change, and Our Future.* Princeton: Princeton University Press, 2000.

Brower, Michael, and Warren Leon. *The Consumer's Guide to Effective Environmental Choices.* New York: Three Rivers Press, 1999.

Brown, Lester R. *Plan B 3.0: Mobilizing to Save Civilization.* New York: W.W. Norton, 2008.

Cherry, Lynne, and Gary Braasch. *How We Know What We Know about Our Changing Climate: Scientists and Kids Explore Global Warming.* Nevada City, CA: Dawn Publications, 2008.

Christianson, Gale E. *Greenhouse: The 200-Year Story of Global Warming.* New York: Walker Publishing, 1999.

Dauncey, Guy, with Patrick Mazza. *Stormy Weather.* Gabriola Island, BC, Canada: New Society Publishers, 2001.

Diamond, Jared. *Collapse: How Societies Choose to Fail or Succeed.* New York: Viking, 2005.

Dow, Kirstin, and Thomas E. Downing. *The Atlas of Climate Change.* Berkeley: University of California Press, 2006.

Economy, Elizabeth C. *The River Runs Black: The Environmental Challenge to China's Future.* Ithaca, NY: Cornell University Press, 2004.

Editors of Collins. *Fragile Earth: Views of a Changing World.* London: Collins, 2006.

Esty, Daniel C., and Andrew S. Winston. *Green to Gold: How Smart Companies Use Environmental Strategy to Innovate, Create Value, and Build Competitive Advantage.* New Haven: Yale University Press, 2006.

Flannery, Tim. *The Weather Makers: How Man Is Changing the Climate and What It Means for Life on Earth.* Melbourne: Text Publishing Co., 2005.

Freese, Barbara. *Coal: A Human History.* Cambridge, MA: Perseus Publishing, 2003.

Gelbspan, Ross. *Boiling Point: How Politicians, Big Oil and Coal, Journalists, and Activists are Fueling the Climate Crisis—and What We Can Do to Avert Disaster.* New York: Basic Books, 2004.

———. *The Heat Is On: The Climate Crisis, the Cover-up, the Prescription.* Reading, MA: Perseus Books, 1998.

Goodstein, Eban. *Fighting for Love in the Century of Extinction: How Passion and Politics Can Stop Global Warming.* Burlington: University of Vermont Press, 2007.

Gore, Al. *An Inconvenient Truth.* Emmaus, PA: Rodale, 2006.

Graedel, Thomas E., and Paul J. Crutzen. *Atmosphere, Climate, and Change.* New York: Scientific American Library/HPHLP, 1995.

Hawken, Paul, Amory Lovins, and Hunter Lovins. *Natural Capitalism: Creating the Next Industrial Revolution.* New York: Little, Brown, 1999.

Houghton, John. *Global Warming: The Complete Briefing.* 2d ed. Cambridge: Cambridge University Press, 1997.

Kolbert, Elizabeth. *Field Notes from a Catastrophe: Man, Na-*

ture, and Climate Change. New York: Bloomsbury Publishing, 2006.

Krupnik, Igor, and Dyanna Jolly, eds. *The Earth Is Faster Now: Indigenous Observations of Arctic Environmental Change.* Fairbanks, AK: ARCUS, 2002.

Kunstler, James Howard. *The Long Emergency: Surviving the End of Oil, Climate Change, and Other Converging Catastrophes of the Twenty-first Century.* New York: Atlantic Monthly Press, 2005.

Kunzig, Robert, and Wallace Broecker. *Fixing Climate: The Story of Climate Science—and How to Stop Global Warming.* London: Profile Books, 2008.

Lovejoy, Thomas E., and Lee Hannah, eds. *Climate Change and Biodiversity.* New Haven: Yale University Press, 2005.

Lovins, Amory B., et al. *Winning the Oil End Game.* Snowmass, CO: Rocky Mountain Institute, 2004.

Lynas, Mark. *High Tide: The Truth about Our Climate Crisis.* New York: Picador, 2004.

McDonough, William, and Michael Braungart. *Cradle to Cradle: Remaking the Way We Make Things.* New York: North Point Press, 2002.

Millennium Ecosystem Assessment. *Ecosystems and Human Well-being: Synthesis.* Washington, DC: Island Press, 2005.

Moser, Susanne C., and Lisa Dilling. *Creating a Climate for Change: Communicating Climate Change and Facilitating Social Change.* Cambridge: Cambridge University Press, 2006.

Motavalli, Jim, ed. *Feeling the Heat.* New York: Routledge, 2004.

National Assessment Synthesis Team, USGRCP. *Climate Change Impacts on the United States: The Potential Consequences of Climate Variability and Change.* Cambridge, UK: Cambridge University Press, 2000.

Pollack, Henry N. *Uncertain Science . . . Uncertain World.* Cambridge, UK: Cambridge University Press, 2005.

Register, Richard. *Ecocities: Rebuilding Cities in Balance with Nature.* Rev. ed. Gabriola Island, BC, Canada: New Society Publishers, 2006.

Schneider, Stephen H., and Terry L. Root, eds. *Wildlife Responses to Climate Change.* Washington, DC: Island Press, 2002.

Speth, James Gustave. *Red Sky at Morning: America and the Crisis of the Global Environment.* New Haven: Yale University Press, 2005.

Weart, Spencer R. *The Discovery of Global Warming.* Cambridge, MA: Harvard University Press, 2003.

TOPICAL GUIDE TO INTERNET RESOURCES

Websites come and go, and information changes. The selected sites listed here have broad scope, apparent permanence, and a focus on facts and action. Also watch journalistic sites like *New York Times,* Environmental News Service, and *Grist* magazine. The following lists will be maintained and updated on this book's websites, www.earthunderfire.com and www.worldviewofglobalwarming.org.

THE SCIENCE OF CLIMATE CHANGE

Carbon Trends by the Carbon Project, www.globalcarbon project.org/carbontrends/index.htm.

Department for Environment, Food, and Rural Affairs, UK, www.defra.gov.uk/environment/climatechange. British government website on climate change.

Espere (Environmental Science Published for Everybody Round the Earth), www.atmosphere.mpg.de/enid/87278a1097ec1c167e3a512cdb1ea3b6,0/148.html. Offers a "Climate Encyclopedia" for nonscientists.

Factsheets on Climate and Energy from the University of Michigan, http://css.snre.umich.edu/facts/factsheets.html.

Goddard Institute for Space Studies, NASA, www.giss.nasa.gov/research/news. Research reports and news releases issued by the Goddard Institute.

National Academies Press, http://books.nap.edu/collections/global_warming/index.html. Publications on global warming and climate change.

National Snow and Ice Data Center, http://nsidc.org/sotc. Provides an overview of the status of snow and ice as indicators of climate change.

Natural Resources Defense Council, www.nrdc.org/globalWarming/fgwscience.asp. "Global Warming Science: An Annotated Bibliography" includes seven years of peer-reviewed climate science.

Potsdam Institute for Climate Impact Research, www.pik-potsdam.de/~stefan. Website of a German ocean physicist, presenting a "Climate Change Fact Sheet" and other information on the role of oceans in climate change.

Real Climate, www.realclimate.org. Discussions by climate scientists of their work and responses to it.

Scientific Assessment of the Effects of Global Change on the United States, www.ostp.gov/cs/nstc.

Union of Concerned Scientists, www.ucsusa.org/global_warming/science.

United Nations Educational, Scientific, and Cultural Organization, http://ioc.unesco.org/iocweb/climateChange.php. Portal to climate information.

United Nations Intergovernmental Panel on Climate Change, www.ipcc.ch. This site provides access to the official IPCC Assessment Reports.

US Climate Change Science Program, www.climatescience

.gov. A sister site to the US Global Change Research Program that integrates federal research on global change and climate change.

US Environmental Protection Agency, www.epa.gov/climate change/science/index.html. Scientific information and data on climate change and projections.

US Global Change Research Program, www.usgcrp.gov. A source of federal global change research.

Weart, Spencer. "The Discovery of Global Warming," www.aip.org/history/climate.

REPORTS AND PLANS OF ACTION

American Solar Energy Society, "Tackling Climate Change in the U.S.," www.ases.org/climatechange.

Clinton Global Initiative, "Areas of Focus: Energy and Climate Change," www.clintonglobalinitiative.org/NET COMMUNITY/Page.aspx?&pid=375&srcid=346.

Daily Kos, "Energize America," a grassroots strategic energy plan, www.dailykos.com/storyonly/2006/5/18/62733/6577.

Europa, European Union's strategy for climate change, http://ec.europa.eu/environment/climat/future_action.htm.

Greenpeace, "Energy Revolution: A Sustainable Pathway to a Clean Energy Future for Europe," www.greenpeace.org/international/press/reports/energy-revolution-a-sustainab.pdf.

Institute for Public Policy Research, UK, "Meeting the Climate Challenge," authored by the International Climate Change Taskforce, www.ippr.org.uk/publicationsand reports/publication.asp?id=246.

National Commission on Energy Policy, "Ending the Energy Stalemate," report of a three-year effort to develop consensus recommendations for US energy policy, www.energycommission.org/site/page.php?report=13.

Pew Center on Global Climate Change, "Agenda for Climate Action," www.pewclimate.org/global-warming-in-depth/all_reports/agenda_for_climate_action/index.cfm.

Princeton Environmental Institute, Carbon Mitigation Initiative, a joint project of Princeton University, BP, and Ford Motor Company (including the "Stabilization Wedge" approach), www.princeton.edu/~cmi.

Rocky Mountain Institute, proposals for market-based, profitable measures to reduce greenhouse emissions, www.rmi.org.

Set America Free Coalition, "Blueprint for US Energy Security," www.setamericafree.org.

"Smarter Living," a Swiss proposal for sustainable development centered on the "2000-watt society," www.novatlantis.ch/pdf/leichterleben_eng.pdf.

Stern Review, report on the economics of climate change, www.hm-treasury.gov.uk/independent_reviews/stern_review_economics_climate_change/stern_review_report.cfm.

United Nations Foundation and Sigma Xi, "Confronting Climate Change: Avoiding the Unmanageable and Managing the Unavoidable," http://www.unfoundation.org/SEG.

US Department of Energy, Office of Energy Efficiency and Renewable Energy, "Scenarios for a Clean Energy Future," www.ornl.gov/sci/eere/cef.

US Environmental Protection Agency, climate policy and actions, www.epa.gov/climatechange/policy/index.html.

World Energy Council, "Energy and Climate Change," www.worldenergy.org/publications/124.asp.

Worldwatch Institute and Center for American Progress, "The Renewable Path to Energy Security," a report of the American Energy Initiative, www.americanenergynow.org.

ORGANIZATIONS AND PROGRAMS DEVOTED TO SUSTAINABILITY

American Association for the Advancement of Science, http://www.sustainabilityscience.org.

Apollo Alliance, www.apolloalliance.org. A joint project of business, labor, environmental organizations, and local communities dedicated to creating sustainable, clean energy and jobs.

Building Technologies Program, US Department of Energy, www.eere.energy.gov/buildings. Technologies that support energy efficiency and renewable energy.

Ceres, www.ceres.org. A coalition of investors and environmentalists devoted to sustainable prosperity.

Climate Biz, www.climatebiz.com. Offers environmentally responsible resources for businesses of all sizes.

Earth Policy Institute, www.earth-policy.org.

Energy Star, US Environmental Protection Agency and US Department of Energy, www.energystar.gov. A government-backed program promoting energy efficiency in homes and businesses.

Environmental and Energy Study Institute, www.eesi.org/index.html. Information and research on energy alternatives and solutions.

ICLEI—Local Governments for Sustainability, www.iclei.org.

United Nations Global Compact, http://www.unglobalcompact.org/AboutTheGC/index.html. Principles of global citizenship and sustainability.

World Business Council for Sustainable Development, www.wbcsd.org. See especially their "Pathways to 2050."

World Resources Institute, www.wri.org.

World Watch, www.worldwatch.org.

DOING OUR PART

American Council for an Energy-Efficient Economy, www.aceee.org/consumer/consumer.htm. Tips on home energy savings and motor vehicle efficiency.

Carbon Monitoring for Action, http://carma.org/.

Climate Change Education, www.climatechangeeducation.org. Links to museums, lesson plans, and other resources relating to climate change.

Climate Institute, "What You Can Do," www.climate.org/topics/you/index.shtml. Recommendations for reducing our environmental footprint.

Earth Day Network, "Climate Change Solutions," www.earthday.net/resources/2006materials/default.aspx.

Focus the Nation, www.focusthenation.org. Nationwide college organizing committee on global warming solutions and political awareness.

Interfaith Power and Light, www.theregenerationproject.org.

National Council of Churches of Christ, www.protecting creation.org. Interfaith campaign for climate justice.

350.org, www.350.org.

Vanity Fair, "Fifty Ways to Help Save the Planet," www.vanityfair.com/politics/features/2006/05/savetheplanet 200605.

Weather Channel, "Solutions Library," http://climate.weather .com.

World Council of Churches, "Ecumenical Earth" site, www .wcc-coe.org/wcc/what/jpc/ecology.html. Resources on climate change.

GOING "CARBON NEUTRAL" WITH CARBON OFFSETS

Clean Air–Cool Planet, "A Consumer's Guide to Retail Carbon Offset Providers," www.cleanair-coolplanet.org/ConsumersGuidetoCarbonOffsets.pdf.

"Climate Credits," www.nature.com/news/2006/061218/full/444976a.html. An article from *Nature* (available for purchase).

Climate Trust, "Carbon Counter," www.carboncounter.org/offset-your-emissions/personal-calculator.aspx. One way to inventory your carbon emissions.

David Suzuki Foundation, "How to Go Carbon Neutral," www.davidsuzuki.org/Climate_Change/What_You_Can_Do/carbon_neutral_steps.asp.

Friends of the Earth, Greenpeace, and WWF-UK, "Joint Statement on Offsetting Carbon Emissions," www .agreenerfestival.com/WWF-GP-FoE_on_offseting.pdf.

Friends of the Earth, Greenpeace, and the World Wildlife Fund warn against tree planting schemes.

Gold Standard, www.cdmgoldstandard.org.

Green-e, www.green-e.org.

Grist: Environmental News and Commentary, "On Your Mark, Offset, Go: A Guide to Offsetting Your Carbon Emissions," www.grist.org/news/maindish/2006/10/10/gies.

DISCUSSION AND INSPIRATION

Bruce Mau Design and the Institute without Boundaries, www.massivechange.com/category/energy. A project devoted to improving the welfare of humanity, including technological solutions to climate change.

The Earth Charter, www.earthcharter.org. A declaration of fundamental principles for building a just, sustainable, and peaceful global society for the twenty-first century.

"A Few Things Ill Considered," http://illconsidered.blogspot .com/2006/02/how-to-talk-to-global-warming-sceptic .html. Frequently heard critiques of global warming, with replies and references.

"An Inconvenient Truth," www.climatecrisis.org. Focusing on the movie *An Inconvenient Truth* and other sources of information on climate change.

Natural Capital Institute, www.naturalcapital.org. Home of the WISER (World Index for Social and Environmental Responsibility) projects, including WiserEarth and Wiser-Business.

Open Democracy, www.opendemocracy.net/climate_change/index.jsp;jsessionid=20EF7E329D61EE5862A06F6EEC446 CD2. Thoughtful essays on the implications of climate change.

Rockefeller Brothers Fund, US in the World, "Talking Global Issues with Americans: A Practical Guide," www.usinthe world.org.

Tom Bender, www.tombender.org. The website of architect and economist Tom Bender, whose "Factor 10" economic principles have been endorsed by the European Union, the World Business Council for Sustainable Development, and the United Nations Environment Programme.

World Changing, www.worldchanging.com. A site devoted to exploring the tools, models, and ideas for building a better future.

Yale School of Forestry and Environmental Studies, Project on Climate Change, http://environment.yale.edu/climate. Also see their on-line magazine at http://e360.yale.edu.

ILLUSTRATION CREDITS

p. iii   1917 photograph by A. O. Wheeler, © Library and Archives Canada. Reproduced with the permission of the Minister of Public Works and Government Services Canada (2006).

p. 12   Satellite image by NASA, JPL, MISR team.

p. 21   Satellite photographs from NASA.

p. 24   Illustration by Jack Cook, Woods Hole Oceanographic Institution.

p. 34   1936 photograph by Erwin Schneider, archives of the Österreichischer Alpenverein, Innsbruck, Austria.

p. 37   Photograph © Lonnie Thompson.

p. 39   Photograph by Vladimir Mikhalenko, courtesy Lonnie Thompson.

p. 40   Satellite photograph adapted from J. Kargel and R. Wessels, GLIMS/USGS/NASA.

p. 42   1914 photograph from the National Oceanic and Atmospheric Administration Photo Library.

p. 43   Archival photograph by G. Jägermayer, Fa. Würthle, Salzburg, 1875. Collection H. Slupetzky, Salzburg, Austria.

p. 44   1859 etching by Eugene Ciceri, courtesy Stefan Wagner, from *Über die Furka* (Beselich: Kissel Verlag), 1999.

p. 63   Photograph courtesy of Alaska Department of Natural Resources, Division of Forestry.

p. 104   Illustration adapted from R. P. Neilson and R. J. Drapek, "Potentially Complex Biosphere Responses to Transient Global Warming," *Global Change Biology* 4 (1998): 505–21.

p. 124   Satellite photograph from NASA, Goddard Space Flight Center, Scientific Visualization Studio.

p. 137   Photograph © Brendan Bannon, Office of the United Nations High Commissioner for Refugees.

p. 141   Photograph © 2006 Nick Graham.

p. 143   Photograph © 2005 Nick Graham.

p. 146   Photograph courtesy Dr. Caroline Rogers, USGS.

p. 163   Illustration adapted from David Talbot, "CO_2 and the Ornery Climate Beast," *Technology Review* (July/August 2006): 40–41. Used by permission.

p. 166   Illustration adapted from *Navigating the Numbers*, World Resources Institute, 2005.

p. 167   Illustration adapted from Hans Rosling, Gapminder Foundation, www.gapminder.org. Data for year 1999 from World Development Indicators, World Bank 2003.

p. 206   Adapted from Robert Socolow and Stephen Pacala, "The Urgency of Carbon Mitigation," presentation given at the DEFRA book launch for Hans Joachim Schellnhuber et al., eds., *Avoiding Dangerous Climate Change*, May 9–10, 2006, Washington DC.

p. 209   Map courtesy of GLOBIO/United Nations Environment Programme. Used by permission.

in, 65, 91, 92; glaciers receding in, *iii,* *27, 32,* 72; ice shelves disintegrating in, 22; military operations by, 66; Native ways of life in, 70–75, 196; warming in, 22, 60, 92; West Nile virus in, 134

Cap-and-trade programs, 178, 197, 199, 207, 214

Cape Hatteras, 124, *127*

Carbon: and greenhouse gases, 242n11; stored in coral reefs, 140; stored in forests, 64, 83–84, 93; stored in oceans, 103

Carbon dioxide: airplanes as source of, 201; automobiles as source of, 167, *168,* 196, 201; coal burning as source of, 171–72, *173,* 214; emissions tabulated by country, *167;* food crops impacted by, 139; forest fires' release of, 64, 93; forests' absorption of, 83–84, 93, 138; forests' release of, 83–84; as greenhouse gas, 7, 10, 52, 53, 73, 114, 131, 145, 242n11; heat wave as source of, 139; in interglacial periods, 21; legal rulings on, 196; oceanic acidity caused by, 103, 108, 114, 145–46, 147, *148;* oceanic biota's absorption of, 108, 147; ocean's absorption of, 144–45; permafrost thawing as source of, 24, 52; reduction of, 165, 171, 176–78, 192, 193; respiratory illness correlated with, 134, 135; rising levels of, 7, 8, 10, 21, 64, *85,* 103, 108, 131, 134, 135, 139, 160, 161, *162–63,* 177; sequestration of, 171, 214; tundra warming as source of, 53

Carbon Disclosure Project (CDP), 199

Carbon trading and markets. *See* Cap-and-trade programs

Caribbean Sea, 94, 140, 143, *146,* 147, 162

Caribou, 53, *54,* 58, *59,* 60, *76–77,* 229n14

Carson, California, *xvi–xvii, 168*

Cascade Mountains, *28–29,* 32, 91

Cattle, 165, 172

Ceres coalition, 199

Chagos Islands, *141*

Chapin, Terry, 53

Chernobyl, 178, 179

Chesapeake Bay, 126, *128*

Chevron corporation, 199

Chicago: green roofs in, *194,* 195; heat wave in, *xxiv–xxv*

Chicago Climate Exchange, 199–200

China: climatic warming in, 92, 153; coal deposits in, 171; coal-fired power plants in, 153, *155,* 156, 172, 184; crop yields decreasing in, 139, 150; dams in, 157, 184, 246n40; drought in, 149, 150, *151,* 153, 162; energy and environmental policy in, 157; energy consumption in, 149, 156–57; glaciers receding in, 38, 40, *41,* 92, 153; hydropower in, 184; industrial development in, 149, 153, *154, 155,* 156–57; and Kyoto Protocol, 157, 177, 200; Olympics in, 153, *154;* pollution in, 149, *152,* 153, *154, 155,* 156–57; renewable energy in, 157; social inequality in, 149; solar power in, 184; urbanization in, 149, 191; and U.S.-sponsored climate plan, 176

Chlorofluorocarbons, 7, 177, 200

Chlorophyll, 147

Cholera, 115, 135

Christy, John, 178

Climate, defined, 7

Climate change: defined, 7; economic impact of, 129, 131; extinction of species caused by, 78–79, 105; as factor in social collapse, 15; loss of biodiversity caused by, 78–81, *82,* 83–85, 103, 105, 107, 108; political debate on, 8, 10, 172, 176, 209; predicted impact of, 10, 160, 162; rapidity of, 20, 24–25, 27, 224n16; scientific assessment of, 10, 78–79, 102, 104, 160–61, 207; and water supply, 115. *See also* Ecosystemic changes; Global warming; Human activity

Clinton, Bill, 242n6

Cloud forests, 78–81

Coal, 7, 147, 149, 153, *155,* 156, 157, 165, *165,* 169, *170;* carbon emissions from, 171–72, *173,* 214; and coal-fired power plants, 147, 153, *155,* 172, *173,* 184

Coastal regions: ecosystemic change in,

94, *95,* 96–98; erosion of, 1, *59,* 60, *67,* 68, *85*

Cod, Arctic, 55

Cogeneration, 179

Colombia, 134, 143

Colville River, 56, *59*

Compact fluorescent (CF) light bulbs, 204, 216

Contraction-and-convergence model, 177–78

Contrails, 132, 201

Cooling, climatic: in Antarctica, 11; caused by aerosols, 7, 8; caused by ocean currents, 25; core samples as record of, 38; in Greenland, 15, 27, 36; and Little Ice Age, 7, 15, 27, 36; in mid-twentieth century, 175, 234n1

Cooperative Institute for Research in Environmental Sciences, 16

Coral reefs, *85,* 94, 103, *106,* 117, 140, *141, 142,* 143–47, *148,* 149, 162

Core samples, 6, 8, 9, 12, 18, 25, 26, 35–36, 105, 224n16

Cormorants, *95,* 96

Corporations, carbon reduction efforts by, 198–201

Costa Rica: amphibians in, 78, *79,* 80; coral reefs in, 143; tropical forests in, 78, *79,* 80–81, 84, 109, 111

Crabs, 96, 126, 236n23

Crick, Humphrey, 89

Cuba, 143

Cultural diversity, 108

Currents: lake, 78; ocean, 24–26, 96, 162, 225n27

Curry, Ruth, 25–26

Cycles: Arctic Oscillation, 61; climatic, 7, 22, 36, 136, 138, 224n16, 227n43; of ice formation, 6, 22, 36; Pacific Decadal Oscillation (PDO), 32; water, 7, 8, 25, 61, 79, 108, *110,* 114, 115; weather, 20, 22. *See also* Feedback loops

Cyclones, 117, 119, 125

D

Dai, Aiguo, 131, 132

Dams: artificial, 115, *116,* 157, 165, 180, 184, 246n40; glacial, 27, 35

Netherlands, dikes in, 121–22, *164*

Nevada, nuclear waste in, 179

New Guinea, 109, 119

New Orleans, 122, 124, 125

New York City, energy conservation in, 191, 195

New York state, renewable energy in, 182, 184

New Zealand, 118; glaciers growing in, 226n34

Nickels, Greg, 195

Nile River, 119

Nitrogen dioxide, 153, 200, 201

Nitrogen oxide, 171

Nitrous oxide, 7, 8, 165

NOAA. *See* U.S. National Oceanic and Atmospheric Agency (NOAA)

Nordhaus, William, 205

Norgay, Tenzing, 38

Northwest Passage, 66

Norway, glaciers growing in, 226n34

Nuclear energy, 169, 178–79, 180

Nunavut, 70

O

Ob River, 22

Oceans: acidity of, 103, *106*, 108, 114, 145–46; carbon dioxide absorbed by, 108, 144–45; carbon stored in, 103; circulation patterns in, 144–45

Oceans, warming of, 8, 45, *85*, 96, 97, *97*, 98, 114, 135, *142*, 143, 144–45, 235n2, 241n3; Arctic, 22, 55; Atlantic, 98, 125; Indian, 136; Pacific, 96, 97, 125, 138

Ocean thermal energy conversion, 179, 184

Oechel, Walt, 52–53

Oil: burning of, 7, 149; consumption of, 149, 156, 169, 198; crisis, 177; drilling for, 58, *59*, 60, 66; and greenhouse-gas emissions, 165, *166*; importation of, 156, 168, 190, 210; production of, 169; transportation system's dependence on, 167, *168*, 171

Orchids, 81, 111

Oregon: coastal ecosystems in, *95*, 96; energy conservation in, *193*, 195; glacial recession in, *28–29*; public

transit in, *193;* seasonal effects of warming in, 86; solar power in, 182

Organic food, *203*, 204

Organisation for Economic Co-operation and Development (OECD), 121

Osterkamp, Tom, 22, 23, 24, 65–66

Overfishing, 84, 98, 107, 109, 140

Overpeck, Jonathan, 21, 72–73

Ovshinsky, Stan, 182

Owls, 94

Oxygen: produced by forests, 61, 64; produced by phytoplankton, 146

Ozone, 7, 139, 165; depletion of, 6, 175, 177, 200, 235n1

P

Pacala, Stephen, *206*, 207

Pacific Decadal Oscillation (PDO), 32

Pacific Ocean: tidal flooding of islands in, *112–13*, 116–19, *117;* warming of, 96, 97, 125, 138

Palmer, Nathaniel B., 14

Palmer Station, 14–15, 48

Pamir Mountains, 38

Panama, 143

Pangnirtung, 70–74

Parks. *See* Protected areas

Parmesan, Camille, 98, *99*, 100–102, 105, 107, 209

Pasterze Glacier, 30, *31, 43*

Pataki, George E., 197

Patz, Jonathan, 134, 136

Paul, Frank, 32

Paulay, Gustav, 140

Pauli, Harald, *90, 91*

Peak production of oil and natural gas, 169, 172

Pearce, Fred, 191

Peat, *23*, 52, 64, 93, 186, 228n4

Penguins, 9, 48, *49*, 51, 52, *85*

Pennsylvania, nuclear accident in, 179

Penny Ice Cap, 72

Pentagon study of climate change, 25

Permafrost, 9, 22, *23*, 24, 52, 64, 65–66, 68, *69*, 72, 73, 74, 153, 162

Peru: glaciers receding in, 33–35, *34, 37;* tropical forests in, 80, 83, *110*

Pesticides, 84, 139, *202*

Peters, Robert L., 103

Peterson, Bill T., 96

Peterson, Bruce, 22

Petroleum. *See* Oil

Pew Center on Climate Change, 207

Philippines, 186

Photosynthesis, 84, 93, 108, 139, 147

Photovoltaic cells, 179, *181*, 182

Pikas, 93–94, 101

Pimentel, David, 188, 203

Pitelka, Louis, 107, 109

Plankton, 15, 25, 78, *85*, *95*, 96, 98, 105, 108, 135, 146–47, 157

Plant life, impact of warming on: in Africa, 78; in Arctic region, 9, 24, 52, 53, *54*, 55, 60, 61, 62–63, 64–65; in Australia, 81; in boreal forests, 61, *62–63*, 64–65; in Costa Rica, 81, 84; in Europe, *viii–ix*, 89, *90*, 91; in Greenland, 9; in mountain environments, *viii–ix*, 27, 34, 35, 78, 89, *90*, 91; in North America, 86; in Peru, 35, 36, *37*, 38, 83; scientific assessment of, 102, 234n34; in tropical forests, 81, 83–84, 111. *See also* Extinction of species; Migration; *names of specific plants*

Plutonium, 179

Polar bears, *vi–vii*, 55–56, *57, 58*, 73, *85*, 229nn12–13

Polar ice sheets, reduction of: in Antarctica, 11, 14, 45, 48; in Greenland, 9, 16–20, 45, 72, 73, *85*; and rising sea level, 13, 14, *19*, 20–21 40, 45, 73

Polar ice shelves, disintegration of: in Antarctica, *5*, *6*, 9, 11–14; in Canada, 22

Policy. *See* Energy policy; Environmental policy

Political activism, 75

Political debate on climate change, 8, 10, 172, 176, 209

Political relations, effect of climate change on, 45, 115, 119, 210, 213

Pollen, 105, 134, 135

Pollination, 107, 108

Pollution: air, 7, 8, 73, *85*, 93, 132, 138, 149, 171; and disease, 94, 153; industrial, 7, 109, 132, 153, *154*, 155, 156–57;

U.S. National Snow and Ice Data Center (NSIDC), 9, 11, 13, 27
Utah, drought in, *133*

V

Venice, tidal flooding in, 122, *123*
Venture capital, 200–201
Vikings, 15
Virgin Islands, 143, *146*
Volcanos, 161, 184, 186, 226n34

W

Walker, David, *88*
Walker Circulation, 138
Wal-Mart corporation, 199, 204
Walrus, 61, 73
Warblers, 81, *88*, 89
Ward Hunt Ice Shelf, 22
Warming, climatic. *See* Atmosphere: warming of; Global warming; Land areas, warming of; Oceans, warming of
Washington state: coastal ecosystems in, 96; glaciers growing in, 226n34
Water cycle, 7, 8, 25, 61, 79, 108, *110*, 114, 115, 144

Water management, 115–16
Water shortage, 30, 33, 38, 115, 139, 150, *151*, 153. *See also* Drought
Water vapor: as greenhouse gas, 7, 131; increase in, 48, 131
Watt-Cloutier, Sheila, 75
Wave and tidal generators, 179, 184
Weart, Spencer, 209
Weddell Sea, 9, 12
Westerling, Tony, 92
West Nile virus, 134, 135
West Virginia, *170, 173,* 180
Whales, *50,* 61
Wildlife Society, 89
Willapa Bay, 126, *128*
Williams, Mark, 149
Williams, Steve, 81, *82, 83*
Wilma, hurricane, 125, 135
Wilson, E. O., 211
Wind power, *xx–xxi,* 171, 179–80, 195, 245n31; birds endangered by generators for, 180, 245n33
Winter temperatures: in Antarctica, 9, 15, 48; in Arctic Ocean, 22; in Arctic region, 9, 16; in Canada, 92
Wisconsin, warming in, 131

Wood stoves, 186
World Bank, 153
World Conservation Monitoring Center, 60
World Health Organization (WHO), 134
World Heritage Sites, 81, 83, *106,* 109, 121, 126, 140, 149
World Meteorological Organization, 125, 175
"World View of Global Warming" project, 1, 2

Y

Yamani, Sheikh Zaki, 206
Yangtze River, 38, 119, 184, 246n40
Yao Tandong, 38
Yellow River, 153
Yohe, Gary, 102
Younger Dryas period, 20, 25, 26
Yucca Mountain, 178, 189
Yukon River, 61

Z

Zemp, Michael, 32
Zeta, tropical storm, *124*
Zimov, Sergey, 52

DESIGNER NICOLE HAYWARD | **COMPOSITOR** INTEGRATED COMPOSITION SYSTEMS | **TEXT** 10.25/14 MINION

DISPLAY DIN | **INDEXER** ANDREW JORON | **PRINTER/BINDER** GOLDEN CUP PRINTING CO., LTD., CHINA